Differentiating Instruction
in Algebra 1

FOR GRADES 7-10

Differentiating Instruction
in Algebra 1

Ready-to-Use Activities for All Students

Kelli Jurek

PRUFROCK PRESS INC.
WACO, TEXAS

Prufrock Press Inc.
P.O. Box 8813
Waco, TX 76714-8813
Phone: (800) 998-2208
Fax: (800) 240-0333
http://www.prufrock.com

Table of Contents

Acknowledgments

Writing a book was not something that I had ever thought about until one of my education professors, Dr. Jacque Melin, of Grand Valley State University, encouraged me to submit my work from a graduate project. I would like to thank her for the confidence she displayed in my work and for helping me to get started. This book would not have become a reality without her.

I would also like to thank my husband, Kevin, and daughters, Erin and Katie, for continually asking me about the progress of my work and encouraging me to keep going. They kept me focused and on task. I would like to thank Benedikte Johnsen, a former exchange student from Norway who is like a daughter, for volunteering to be the first to purchase this book even though she has little use for it.

I would also like to acknowledge several of my very talented colleagues at Sparta Area Schools. A big thanks goes to Sue Koning and Erin Ondrusek, Sparta Middle School math teachers, for reviewing chapters and offering valuable suggestions. In addition, I appreciate Michelle Jonker, Sparta High School English and foreign language teacher, for helping me edit the introduction to the book. It would have been much harder to complete this project if I had not had their help and support.

Introduction

My grandmother inspired me to become an educator. She was a teacher in a one-room schoolhouse in Plymouth, MI, in the 1930s, and she shared many stories of her 20 students who ranged in age from 5 to 18. Her students were not divided into groups based on chronological age; they worked with other students at their same learning level. It was common for more advanced students to assist less-accomplished students, giving my grandmother time to provide instruction and remediation for individuals and small groups of students. Once students could navigate the lessons independently, she was able to move students to the next level. Why do I mention my grandmother? I believe that teachers in the early 20th century used differentiated learning strategies on a daily basis. It was a survival strategy that worked, and we can learn from these early educators.

The idea for this resource stemmed from my frustration with not being able to find ready-to-use differentiated materials for algebra. The most readily available differentiated resources seem to address social studies material or K–8 curricula. However, the mixed-ability classrooms in our schools demand differentiated instruction for all subjects and all grade levels. Ready-to-use materials for middle and high school teachers are hard to find, yet they are necessary for teachers who do not have the time to create tiered activities to help their students achieve success.

How to Use This Book

Differentiating Instruction in Algebra 1 is not a textbook. Before students are assigned the activities in this book, they will likely require some level of direct instruction. This book contains four units covering introduction to functions, systems of linear equations, exponent rules and exponential functions, and quadratic functions. These units are very broad, and it would be impossible to cover all aspects of these topics given space constraints of this book. However, based on my experience, I have chosen lessons for each unit that most closely match what might be taught in Algebra 1 courses.

Each unit begins with launch scenarios and a preassessment that can be used to determine the readiness level of each student. Each lesson within the unit includes both group activities and individual practice pages that enable students to make choices or are tiered to three readiness levels. Each unit ends with a project that enables students to demonstrate the knowledge they have acquired. Grading rubrics are provided within each unit, and an answer key for all four units is included at the end of the book.

In addition to using this resource in a general education classroom, it could also be used by parents who homeschool and need practice work for their children. Educators with gifted and talented students or students receiving special education resources will also benefit from the choice activities and the tiered learning opportunities.

Why Differentiate?

The techniques my grandmother used to teach her students were not called differentiation at that time, but her teaching incorporated strategies that are now considered best practices in education. Differentiation is defined by Tomlinson (2001) as providing different options of acquiring content, of processing or making sense of ideas, and of developing products that enable students to demonstrate their learning. I believe that differentiation includes offering choices to students, tailoring a homework assignment based on the needs of your students, and working with a group of students to customize learning. These are all things that educators do on a daily basis to meet the needs of their students.

Educators know students come to our classrooms with varied levels of understanding, experience, and interest in mathematics. Because of the large number of students taught in secondary classrooms, I do not believe it is practical to think that educators can tailor instruction to fit the varied interests and learning styles of every student. It is practical, however, to offer assignments and small-group activities that challenge students at their readiness level and scaffold them to the next level. Russian psychologist Lev Vygotsky introduced the concept of scaffolding in the early 20th century. Benson (1997) defined scaffolding as a design to support students until they are able to complete tasks independently. Benson noted that scaffolding involves the teacher identifying the skills or information students are missing and then bridging the gap so that students can acquire those skills or information. We, as educators, can build these bridges through the differentiated strategies included in this resource.

Alignment With National Standards

Each unit in *Differentiating Instruction in Algebra 1* contains lessons that are aligned with the Common Core State Standards (CCSS). "The standards were developed in collaboration with teachers, school administrators, and experts, to provide a clear and consistent framework to prepare our children for college and the workforce" (Common Core State Standards Initiative, n.d., para. 1). According to the National Council of Teachers of Mathematics (2010), "states will be asked to adopt the Common Core State Standards in their entirety and include the core in at least 85 percent of the state's standards in English language arts and mathematics" (para. 2).

As more districts adopt the CCSS, it is imperative that teachers review what they are teaching and how their lessons support the new learning standards. One challenge is that the high school mathematics standards are grouped by six categories: number and quantity; algebra; functions; modeling; geometry; and statistics and probability. Similar to previous high school mathematics standards, topics taught in Algebra 1 will be included in numerous categories, and it is sometimes difficult to separate the Algebra 1 and Algebra 2 standards. At the beginning of each unit, I have identified the standards that are addressed or partially addressed in each unit.

Summary of Differentiated Components

Following is a description of each type of differentiated/small-group activity that may be included in each unit. Many of the activities include three levels of work based on student readiness. For students who have not yet mastered the concepts of the lesson, the practice page will be indicated with a moon symbol at the top. Students working on level who are ready to practice the concepts with little assistance will complete the practice page with a sun symbol at the top. Students who have mastered the concepts and need to be challenged will complete the practice page with a star symbol at the top.

Preassessment

The preassessment can be used to determine the readiness level of each student for the lessons within a unit. There is a space for students to identify whether or not they are confident of their answer for each problem. Each question in the preassessment is tied directly to one of the subsequent lessons in the unit.

Learning Targets and Study Guide

The learning targets and study guide handout should be distributed to students at the beginning of the unit. It enables students to see what topics will be covered in the unit and can serve as a review guide. As a concept is taught, students can complete the learning target page with definitions and example problems.

Vocabulary Choice Board

The vocabulary choice board offers students the opportunity to choose how to demonstrate their understanding of the critical vocabulary. Choices include preparing LINCS cards (Ellis, 2000), creating a graphic organizer, writing and delivering a news broadcast, drawing a cartoon strip, writing a creative story, writing and performing a rap, and acting out the words. LINCS (Ellis, 2000) is a strategy that helps students learn and remember critical vocabulary. The acronym LINCS represents: List the parts, Identify a reminding word, Note a LINCing story, Create a LINCing picture, and Self-test. The first step is to list the parts of the LINCS cards, which include the new vocabulary term, the definition, the reminding word, the LINCing story, and the LINCing picture. After the definition is recorded, the reminding word is chosen. It must be a real word that has a meaning that the students already know. The LINCing story should be simple and must include the reminding word and connect it to the new word. The student can self-test by writing the new term and the reminding word on one side of a note card and then writing the definition, LINCing story, and LINCing picture on the other side of the note card. A rubric is included with the vocabulary choice board.

Checkerboard

The checkerboard puzzles encourage students to work in pairs or small groups to solve mathematical problems that review algebra concepts.

Tic-Tac-Toe Boards

These resources provide choices for students and can be used to review concepts or as homework. Students can choose the problems they want to complete. In addition, teachers can assign additional problems for students who need further practice.

Exit Slips

For each unit, there is a series of exit questions that can be used as formative assessments. The exit slip can be given to students approximately 10–15 minutes before the end of the class. Students then complete the questions before they leave the classroom. Teachers can use the responses from the exit slip to guide lesson planning for the next session.

Hexagon Puzzles

Each puzzle page includes 24 practice problems that increase in difficulty from left to right across the page. Students choose a starting point in the far left-hand column and complete the problem. If it is solved correctly, they will choose an adjacent problem in the next column and complete the problem. If it is solved correctly, they will continue this process, moving to the right. If at any time a problem is not solved correctly, the student must choose a different problem in the same column and complete the problem. When they solve a problem correctly, they move to the right. In order to complete the puzzle page, students must have a complete path with adjoining hexagons from the left to the right. Students can either be provided with an answer key to check their answers as they work, or the teacher can check their answers for them. Students may also be assigned all of the problems for more practice.

Agenda

The agendas include three sets of practice problems. The first group is called *Imperatives*, and students complete all of the problems. This set of problems is generally a review of basic concepts. The second section is called *Negotiables*, which generally includes three problems of varying difficulty. Students select at least two problems to complete based on their interest. The last section is *Options*, which includes an optional, more challenging problem. Depending on each student's progress, the number of problems assigned can be varied.

Structured Academic Controversy

This activity enables students to discuss and support both sides of a mathematical argument (Johnson, Johnson, & Smith, 2000).

Projects

Several of the units include a culminating project that can be used as an assessment or be assigned to students who are working above level as a challenge problem.

RAFT Activity

This activity enables students to demonstrate their understanding of a specific concept through a writing assignment. Students choose a role, an audience, a format, and a topic according to their interests (Santa, 1988). Students may also propose a role, audience, format, and topic that match their personal interests. A rubric is provided.

TriMind Activity

Students choose from three activities based on Sternberg's (2010) creative, analytical, and practical thinking styles. A rubric is provided.

Think Dots

This tiered activity allows students to work in small groups to solve six problems. The problems can be done in any order, or the order can be determined by rolling a die. If students are working in mixed-ability groups, students may be asked to demonstrate how to complete their problem to their group. Students can also work independently.

Unit 1

Introduction to Functions and Relationships

Beginning algebra students need to take their understanding of linear equations and solving simple one-variable equations and apply it to studying special relations of data called functions. Functions describe everyday situations where one specific quantity determines the value of another. Students must learn to write and evaluate functions because they describe a unique relationship between two quantities and are frequently used to model everyday situations.

This unit begins with a preassessment and three real-life applications of functions that can be discussed in small groups and then as a larger group. Many of the activities will offer the students an opportunity to choose learning activities according to their learning style, personal interests, and readiness level.

What Do We Want Students to Know?

Common Core State Standards Addressed: • 8.F.1, 4 • A.CED.1, 2 • F.IF.1, 2, 5 • F.BF.1c	Big Ideas • Not all relationships are functions. • A function denotes a special relationship between independent and dependent variables. • A function must pass the vertical line test.
	Essential Questions • What makes a relationship a function? • How is the vertical line test used to determine if a relationship represents a function? • Does it matter that a set of data does not represent a function?

Critical Vocabulary

Domain

Function notation

Dependent variable

Output

Range

Relation

Function

Independent variable

Vertical line test

Input

Unit Objectives

As a result of this unit, students will know:

➤ a function is a special type of relation,

➤ the difference between a relation and a function,

➤ all functions are relations but not all relations are functions,

➤ the difference between an independent and dependent variable, and

➤ $f(x)$ is read "f of x" and $f(-1)$ means to evaluate the function rule by substituting -1 into the expression.

As a result of this unit, students will understand that:

➤ in order for a relation to be a function, each domain element is associated with a unique range element;

➤ if a relation is a function, it can be written as an equation using function notation; and

➤ the independent variable is graphed on the horizontal axis, and the dependent variable is graphed on the vertical axis.

As a result of this unit, students will be able to:

➤ use the vertical line test to tell if a relationship is a function,

➤ identify the domain and range of relations and functions,

➤ determine reasonable domains for given situations, and

➤ evaluate functions for a given domain or range.

Launch Scenarios

➤ If we listed 10 cities and their corresponding area codes, would these relations represent a function? (Lesson 1)

➤ A lawyer charges $150 per hour for his services. Identify the dependent and independent variables, and then write the equation to represent the function. (Lesson 2)

➤ Give a real-life situation that can be modeled by the function $h(t) = 5 + 5t$. (Lesson 3)

Unit Overview: Introduction to Functions and Relationships

(Assumes a 50–60-minute class period)

Lesson	Whole-Class/Small-Group Discussion and Activities	Individualized Learning Activities
Preassessment (1/2 period)	Give the preassessment at least several days prior to beginning the unit (20 minutes)	
Lesson 1: Introduction (2 periods)	Share results of the preassessment (20 minutes) Introduce launch scenarios and discuss possible solutions (10 minutes) Hold a classroom discussion on functions (20 minutes)	
		Learning Targets and Study Guide (15 minutes) Vocabulary Choice Board (15 minutes) Agenda (20 minutes) Exit Slip (10–15 minutes)
Lesson 2: Writing Functions (1–2 periods)	Hold a classroom discussion on writing functions (20 minutes)	
		Agenda (30 minutes) Tic-Tac-Toe Board (30 minutes) Exit Slip (10–15 minutes)

Lesson	Whole-Class/Small-Group Discussion and Activities	Individualized Learning Activities
Lesson 3: Evaluating Functions and Function Notation (1–2 periods)	Hold a classroom discussion on evaluating functions and function notation (20 minutes)	
		Agenda (30 minutes) Hexagon Puzzles (20 minutes) Exit Slip (5–10 minutes)
Lesson 4: Wrap Up and Assessment (2 periods)	Collect learning targets and study guide handouts and vocabulary choice board projects	
		Think Dots (30 minutes) Tri-Mind Activity (30 minutes) Exit Slip (15 minutes)

Introduction to Functions and Relationships

Preassessment

Directions: Solve the following problems to the best of your ability, and then rate your confidence in your answer in the space provided. Skip questions you do not know. This will not be graded.

1. Identify the domain and range for the following relations:

   a. $\{(2,4),(1,3),(7,5),(2,5),(4,4)\}$

   b.
x	−2	−1	0	1	2
y	4	3	1	5	−3

   Domain: _____

   Domain: _____

   Range: _____

   Range: _____

   _____ I'm sure

   _____ I'm not sure

2. Circle the relationships that represent functions.

   a.
x	−1	0	1	2	3	4	5
y	2	4	2	5	2	6	9

   b. $\{(2,4),(1,3),(7,5),(2,6),(4,4)\}$

   c. $y = 2x - 4$

   d. $y^2 = 2x^2 + 8$

   _____ I'm sure

   _____ I'm not sure

3. Sketch a graph that represents a function and a graph that does not represent a function.

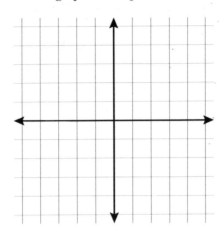 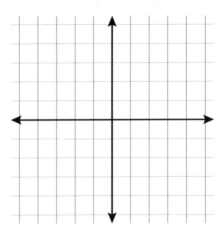

_____ I'm sure

_____ I'm not sure

4. Evaluate the functions for the given variable.

   a. $f(x) = 2x + 7$ for $x = -1$ b. $f(x) = -2x + 7$ for $x = 5$

   c. $g(x) = x^2 - 5$ for $x = 2$ d. $g(x) = 4x - 5$, find x when $g(x) = 0$

   _____ I'm sure

   _____ I'm not sure

5. Using $f(x) = 2x - 1$ and $g(x) = -4x + 4$, simplify the following expressions:

   a. $f(x) + g(x)$

   b. $g(x) - f(x)$

   c. $f(x) - g(x)$

   _____ I'm sure

   _____ I'm not sure

6. Identify the independent and dependent variables and then write the function for the situation described below.

   A plumber charges $50.00 for a service call to your home plus $45.00 per hour.

   _____ I'm sure

   _____ I'm not sure

Introduction to Functions and Relationships
Learning Targets and Study Guide

At the end of this unit, every student should be able to say:

Learning Target	Explain and Give an Example
I can determine the domain and range for a function. Initial here when mastered _____	
I can determine if a table of data represents a function. Initial here when mastered _____	
I can determine if an equation represents a function. Initial here when mastered _____	
I can sketch a graph that is a function and a graph that is not a function. Initial here when mastered _____	
I can explain how to evaluate a function. Initial here when mastered _____	
I can show how to simplify expressions written in function notation. Initial here when mastered _____	
I can identify independent and dependent variables. Initial here when mastered _____	
I can write a function to represent an everyday situation. Initial here when mastered _____	
I can solve a real-life story problem involving functions. Initial here when mastered _____	

Introduction to Functions and Relationships

Vocabulary Choice Board

Directions: Using the critical vocabulary for the unit, choose one of the following methods to demonstrate your knowledge of the mathematical terms.

LINCS Cards	Graphic Organizer	News Broadcast
Prepare a LINCS card for each vocabulary word to demonstrate your understanding of the vocabulary words. *(logical–mathematical)*	Construct a graphic organizer of your choice to define each word and demonstrate your understanding of the vocabulary words and how they fit into the current unit. *(verbal/language–mathematical)*	Write a news broadcast that defines all of the vocabulary words, and provide stories that demonstrate your understanding of the vocabulary words. Turn in your news broadcast, and then present it to the class. *(intrapersonal–language)*
Nature Survival	**Creative Story**	**Poster Board**
Prepare a nature survival guide about how each of the vocabulary words would be used if you were stranded on a deserted island. The guide must demonstrate your understanding of the vocabulary words. *(natural–language)*	Write a creative story with color illustrations that demonstrates your understanding of the vocabulary words. *(verbal–language)*	Prepare an 11" x 14" poster board that contains the definitions of the vocabulary words and includes one real-life example for each word that demonstrates your understanding of the definition. *(language–spatial)*
Cartoon	**Rap or Song**	**Acting Out**
Prepare a cartoon strip with at least five frames that defines each word and is illustrated with appropriate pictures to demonstrate your understanding of the vocabulary words. *(visual–spatial)*	Write a rap or a song containing all of the vocabulary words and their correct definitions. Perform your rap or song for the class. *(musical–language)*	Write a script and then act out the words with their correct definitions. Use props if appropriate. *(kinesthetic–language)*

Critical Vocabulary

Domain	Range	Vertical line test
Function notation	Relation	Input
Dependent variable	Function	
Output	Independent variable	

Introduction to Functions and Relationships
Vocabulary Choice Board Rubric

Grading scale:	3 Well done	2 Close to requirements		1 Needs more work
Choice	**Definitions**	**Illustrations and/ or creativity**	**Neatness**	**Other**
LINCS cards	Definitions demonstrate knowledge of vocabulary. Score _____	The illustrations are appropriate and creative. Score _____	The cards are easy to read. Score _____	Semantic connections between words are demonstrated. Score _____
Graphic organizer	Definitions demonstrate knowledge of vocabulary. Score _____	The graphic organizer creatively displays the definitions. Score _____	The graphic organizer is easy to read. Score _____	The graphic organizer choice is appropriate for this assignment. Score _____
News broadcast	Definitions demonstrate knowledge of vocabulary. Score _____	The news broadcast is creative and holds the attention of the audience. Score _____	The news broadcast is easy to read. Score _____	Good presentation skills are demonstrated. Score _____
Nature survival guide	Definitions demonstrate knowledge of vocabulary. Score _____	The guide contains illustrations that are appropriate. Score _____	The guide is easy to read. Score _____	The final product resembles a survival guide. Score _____
Creative story	Definitions demonstrate knowledge of vocabulary. Score _____	The story contains illustrations that are appropriate. Score _____	The story is easy to read. Score _____	The story is interesting. Score _____
Poster board	Definitions demonstrate knowledge of vocabulary. Score _____	The poster board includes illustrations that are appropriate. Score _____	The poster board is easy to read. Score _____	Real-life examples are included for each vocabulary term. Score _____
Cartoon	Definitions demonstrate knowledge of vocabulary. Score _____	The cartoon includes illustrations that are appropriate. Score _____	The cartoon is easy to read. Score _____	The cartoon contains a minimum of five frames. Score _____
Rap or song	Definitions demonstrate knowledge of vocabulary. Score _____	The rap or song is creative. Score _____	The lyrics are easy to read. Score _____	Good performance skills are demonstrated. Score _____
Acting out	Definitions demonstrate knowledge of vocabulary. Score _____	The script is creative. Score _____	The script is easy to read. Score _____	Good performance skills are demonstrated. Score _____

Total: _____ /12= _____ %

Comments:

Name: _____ Hour/Block: _____ Date: _____

Lesson 1: Introduction

Agenda

Imperatives (You must do all three of these.)

1. Determine whether the following relations are functions. Identify the domain and range for those that are functions.

a.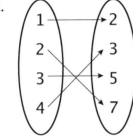

b.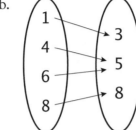

c.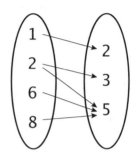

Domain: {   }   Domain: {   }   Domain: {   }

Range: {   }   Range: {   }   Range: {   }

2. Determine whether the following relations are functions. Identify the domain and range for those that are functions.

a. $(-3, -8), (-2, -5), (-3, -2), (0, 1), (1, 4), (2, 7)$

Domain: {   }   Range: {   }

b.

x	-3	-2	-1	0	1	2	3	4	5
y	-2	-5	-3	1	-2	-5	1	-2	-5

Domain: {   }   Range: {   }

c.

x	-3	-2	-1	0	1	2	3	4	5
y	1	1	1	1	1	1	1	1	1

Domain: {   }   Range: {   }

3. Draw two graphs: one that represents a function and one that does not.

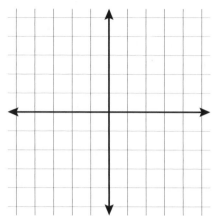 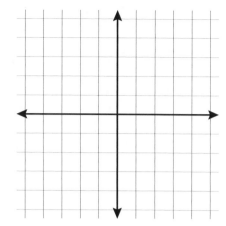

Negotiables (You must do at least two of these.)

1. Determine whether the following relationships represent functions. Explain your answers. If you answer no, give a counterexample.
   a. A person and his or her birth date

   b. First name and last name

   c. City and zip code

   d. Last name and first name

2. Using the previous question as an example, identify two new relationships that are functions and two new relationships that are not functions.

Relationships That Are Functions	Relationships That Are Not Functions

3. Explain how the vertical line test is used to determine whether a relation is a function. Draw two graphs: one that is a function and one that is not a function. Draw the vertical line where appropriate to test each function.

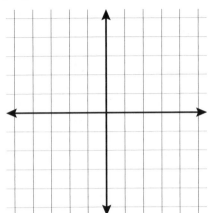 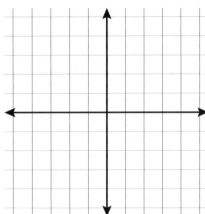

Options (You may choose to do these problems.)

1. Name four different ways to represent a relation or a function.

2. Give an example of a relation that is a function, and represent it in the four ways you listed.

Lesson 1: Introduction

Exit Slip

1. Explain how a function is a special type of relation.

2. Explain how the vertical line test is used to determine if a relation is a function.

3. In the first table, insert the missing numbers so that the table represents a function. In the second table, insert the missing numbers so that the table does not represent a function.

This is a function:

x	−3	−2	−1	0	1		3	4	5
y	0	1	−3	1	−2		1	−2	−5

This is not a function:

x	−3	−2	−1	0	1		3	4	5
y	0	1	−3	1	−2		1	−2	−5

Introduction to Functions and Relationships

Lesson 2: Writing Functions

Agenda

Imperatives (You must do all three of these.)

1. Use the slope formula $m = \dfrac{y_2 - y_1}{x_2 - x_1}$ to determine the slope of the lines containing the following points. Keep your answers as reduced fractions.

   a. $(3,4)$ and $(4,6)$ slope of the line: _____

   b. $(2,-4)$ and $(0,8)$ slope of the line: _____

   c. $(3,-4)$ and $(1,-5)$ slope of the line: _____

   d. $(-2,1)$ and $(-1,7)$ slope of the line: _____

2. Determine the y-intercept from the following tables.

   a.

x	−3	−2	−1	0	1	2	3	4
y	9	7	5	3	1	−1	−3	−5

   y-intercept: _____

   b.

x	−3	−2	−1	0	1	2	3	4
y	−2	−1	0	1	2	3	4·	5

   y-intercept: _____

3. Determine if the following relationships represent functions. If they do, then write the equation that represents the function.

   a.

x	0	1	2	3	4	5	6	7
y	−1	1	3	5	7	9	11	13

   Equation: _____

   b.

x	−3	−2	−1	0	1	2	3	4
y	7	4	1	−2	−5	−8	−11	−14

   Equation: _____

4. Identify the independent and dependent variables in the following situations. An example is shown below.

   An electrician's fee is based on the number of hours worked in your home.

   Rewritten with the word "depends": <u>The fee depends on the hours worked</u>.
   Independent variable: <u>electrician's fee</u>   Dependent variable: <u>hours worked</u>

a. The cost of shipping a package is determined by its weight.

Rewritten with the word "depends": _____

Independent Variable: _____ Dependent Variable:_____

b. The cost of renting a car is based on the number of days it is rented.

Rewritten with the word "depends": _____

Independent Variable: _____ Dependent Variable:_____

c. Apples are sold by the pound.

Rewritten with the word "depends": _____

Independent Variable: _____ Dependent Variable:_____

Negotiables (You must do at least two of these.)

Identify the independent and dependent variables and then write the equation for each situation.

1. An accountant's fee is $150 per hour for services.

   a. Independent Variable: _____ Dependent Variable:_____

   b. Function: _____

2. The cost to get into the carnival is $5.00, and you have to pay $2.00 for each ride.

   a. Independent Variable: _____ Dependent Variable:_____

   b. Function: _____

3. Write functions to represent the following tables.

   a.

x	−5	−3	−1	1	3	5	7	9
$f(x)$	−17	−11	−5	1	7	13	19	25

   Function: _____

   b.

x	−6	−3	−1	2	3	6	8	9
$f(x)$	−1	0.5	1.5	3	3.5	5	6	6.5

   Function: _____

4. Write a function to represent the following situation. Identify the independent and dependent variables.

Katie received $200 from her grandparents on her 10th birthday. She immediately put the money into her large piggy bank. She also receives an allowance of $25 per week for the chores she does around the house. If she puts $20 per week from her allowance into the piggy bank and never takes any money out, how much will be in the piggy bank at any specific time?

a. Independent Variable: _____ Dependent Variable:_____

b. Function: _____

c. How much money will she have in the piggy bank after 3 weeks? _____

d. How much money will she have in the piggy bank after one year? _____

e. How much money will she have in the piggy bank after 5 years? _____

Options (You may choose to do this problem.)

1. Write scenarios that could be represented by the following sets of data or equations. Be sure to identify the independent and dependent variables.

a.

x	1	2	3	4	5	6	7	8
y	2.50	5	7.5	10	12.50	15	17.50	20

b.

x	0	5	8	10	12	15	20	50
y	50	60	66	70	74	80	90	150

c. $y = 2x + 25$

d. $h = 5t + 100$

Name: _____ Hour/Block: _____ Date: _____

Lesson 2: Writing Functions
Tic-Tac-Toe Board

Directions: Choose three problems in a row, in a column, or on a diagonal. Solve the problems and show your work on a separate sheet of paper. Circle the problems you choose on this page and then reference them on your work paper with the number from the square.

1	2	3
Write a function to represent the table of data. 	Write a function to represent each situation: a. The cost of gasoline is $3.59 per gallon. b. Adam's weekly salary is $1,250. c. Soda is $1.89 for a two-liter.	Identify the independent and dependent variables for each function: a. Zach earns $7.50 per hour working at the grocery store. b. Erin's grade reflects the number of hours she studies.

Table for problem 1:

x	y
−1	−8
0	−3
1	2
2	7
3	12
4	17

4	5	6
Write a function to represent the situation described below. Identify the independent and dependent variables. A cell phone plan charges $25 per month to use the phone, plus $0.30 for each text message or phone call.	Write a scenario to represent the function below. Identify the independent and dependent variables. $$f(x) = 75 + 3x$$	Write a function to represent each situation. a. The cost of apples is $1.69 per pound. b. The cost to copy each page is $0.10. c. The health club charges a joining fee of $125 plus $57 per month.

7	8	9
Write a function to represent each situation. Identify the independent and dependent variables. a. The cost of bananas is $0.59 per pound. b. The lawn service company charges $20.00 per week. c. The tutor charges $25.00 per hour.	Identify the independent and dependent variables for each function. a. The cost of a holiday wreath is $1.00 per inch plus $3.00 for a bow. b. The candy costs $1.50 per pound.	Fill in the missing values in the table to represent a linear function.

Table for problem 9:

x	y
−3	37
−1	19
	1
4	−26
7	
8	−62

Introduction to Functions and Relationships

Lesson 2: Writing Functions

Exit Slip

1. Describe how you determine the independent and dependent variables.

2. Describe a situation that can be modeled with a function. Write an equation to model the situation.

Lesson 3: Evaluating Functions and Function Notation

Agenda

Imperatives (You must do all three of these.)

1. Evaluate each function for the given input values.

    a. For $f(x) = -2x + 1$, find $f(x)$ when $x = 1$

    b. For $f(x) = 2x - 5$, find $f(x)$ when $x = 0$

    c. For $g(x) = 6x$, find $g(x)$ when $x = -2$

    d. For $h(t) = -.75t + 12$, find $h(t)$ when $t = 12$

2. Evaluate the function for the given values.

    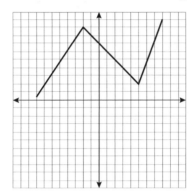

    a. $f(0)$ _____

    b. $f(-2)$ _____

    c. $f(-4)$ _____

    d. $f(5)$ _____

    e. x, when $f(x) = 2$ _____

    f. x, when $f(x) = 1$ _____

3. The function $h(t) = 4 + 8t$ gives the height of a balloon off the ground (in feet) t seconds after it is released. Use this function to answer the questions below.

    a. Find $h(3)$. What is the real-life meaning of this answer?

    b. Find $h(10)$. What is the real-life meaning of this answer?

    c. When is the balloon 76 feet in the air?

Negotiables (You must do at least two of these.)

1. Write a function to represent each of the following situations.
    a. Shelbie earns extra money helping students apply for jobs by preparing resumes and cover letters. She charges $35 plus $0.25 per word.

    b. José must pay the first $1,500 of his medical expenses plus 10% of the remaining charges.

2. Given $f(x) = -4x + 7$ and $g(x) = 3x - 1$,

    a. find $f(x) + g(x)$

    b. find $g(x) - f(x)$

    c. find $2g(x)$

    d. find $3f(x) + 2g(x)$

3. Given $f(x) = -4x + 7$ and $g(x) = 3x - 1$,

    a. is $f(0) = 0$?

    b. is $g(0) = f(2)$?

    c. is $f(-1) = 10$?

    d. is $f(1) + g(2) = g(3)$?

Options (You may choose to do this problem.)

1. Use function notation to write the following:

    a. "f of x is 2 less than 5 times a number"

    b. "f of x is 4 more than 3 times x squared"

Lesson 3: Evaluating Functions and Function Notation

Hexagon Puzzle

Directions: Solve any problem in a hexagon in the far left-hand column. If you get the answer correct, choose a problem from the next column that is adjacent to the first problem. If you get a problem wrong, move to another problem in the same column adjacent to the hexagon and solve it. Make a path to the right by solving the problems. You must complete a continuous path of problems from left to right.

A→ Identify the domain for the coordinate pairs.
$(2,3), (-7,6), (5,7), (-2,5)$

B→ Identify the range for the coordinate pairs.
$(2,3), (-7,6), (5,7), (-2,5)$

C→ Put the data into a table.
$(-2,3), (-3,6), (-5,7), (-8,9)$

D→ Make a mapping diagram to represent the following.
$(1,2), (4,3), (1,7), (6,5)$

E→ Does the relationship between cities and zip codes represent a function?

F→ Does the relationship between a state and its governor represent a function?

G→ Does the relationship between a person and a social security number represent a function?

H→ Does a person and his or her birth date represent a function?

I→ Find the slope of the line that contains the following points.
$(2,3), (5,4)$

J→ Find the slope of the line that contains the following points.
$(1,3), (4,5)$

K→ Find the slope and y-intercept for the line.
$f(x) = 3x + 6$

L→ Find the slope and y-intercept for the line.
$f(x) = -x - 5$

M→ Identify the dependent and independent variables: apples sold by the pound.

N→ Identify the dependent and independent variables: total pay and hours worked.

O→ Identify the dependent and independent variables: babysitting fees and time.

P→ Identify the dependent and independent variables: cost and tickets purchased.

Q→ Given $f(x) = 3x - 2$, evaluate $f(-5)$.

R→ Given $g(x) = x + 12$, evaluate $g(-2)$.

S→ Given $h(t) = 8t - 2$, evaluate $h(0)$.

T→ Given $f(x) = 4x + 7$, evaluate $f(-8)$.

U→ Given $f(x) = x - 3$ and $g(x) = x + 9$, find $f(x) + g(x)$.

V→ Given $f(x) = 2x + 1$ and $g(x) = 3x + 3$, find $f(x) - g(x)$.

W→ Given $g(x) = 5x + 1$ and $h(x) = 6x - 8$, find $g(x) - h(x)$.

X→ Given $g(x) = 4x + 7$ and $f(x) = x - 5$, find $f(x) - g(x)$.

Introduction to Functions and Relationships

Lesson 3: Evaluating Functions and Function Notation

Hexagon Puzzle

Directions: Solve any problem in a hexagon in the far left-hand column. If you get the answer correct, choose a problem from the next column that is adjacent to the first problem. If you get a problem wrong, move to another problem in the same column adjacent to the hexagon and solve it. Make a path to the right by solving the problems. You must complete a continuous path of problems from left to right.

A→Explain the difference between domain and range.

B→Explain how to use the vertical line test. Draw an example of a function.

C→Name the five ways to represent a function.

D→Explain how inputs and outputs are related to domain and range.

E→Does the relationship between cities and area codes represent a function?

F→Does the relationship between a state and its U.S. Senators represent a function?

G→Does the relationship between last name and first name represent a function?

H→Does a person and his or her birth date represent a function?

I→Find the slope of the line that contains the following points. $(2,-3),(5,-4)$

J→Find the slope of the line that contains the following points. $(-1,3),(-5,-4)$

K→Find the slope and y-intercept for the line. $f(t) = 3t - 2$

L→Find the slope and y-intercept for the line. $f(x) = 12 - 3x$

M→Create a function and then identify the dependent and independent variables.

N→Identify the dependent and independent variables: height of the tree and the total cost.

O→Identify the dependent and independent variables: babysitting fees and number of children.

P→Identify the dependent and independent variables: hours of skiing and cost.

Q→Given $f(x) = -3x - 2$, evaluate $f(-5)$.

R→Given $g(x) = -x + 12$, evaluate $g(-2)$.

S→Given $h(t) = -8t - 2$, evaluate $h(0)$.

T→Given $f(x) = -4x + 7$, evaluate $f(-8)$.

U→Given $f(x) = -3x - 2$ and $g(x) = -x + 9$, find $f(x) + g(x)$.

V→Given $f(x) = -2x - 1$ and $g(x) = -3x - 3$, find $f(x) - g(x)$.

W→Given $g(x) = -5x + 1$ and $h(x) = -6x - 8$, find $g(x) - h(x)$.

X→Given $g(x) = -4x + 7$ and $f(x) = -x - 5$, find $f(x) - g(x)$.

Introduction to Functions and Relationships

Lesson 3: Evaluating Functions and Function Notation

Hexagon Puzzle

Directions: Solve any problem in a hexagon in the far left-hand column. If you get the answer correct, choose a problem from the next column that is adjacent to the first problem. If you get a problem wrong, move to another problem in the same column adjacent to the hexagon and solve it. Make a path to the right by solving the problems. You must complete a continuous path of problems from left to right.

A→ Does the relationship between a state and its governor represent a function?

B→ Does a person and his or her birth date represent a function?

C→ Does the relationship between cities and zip codes represent a function?

D→ Does the relationship between last name and first names represent a function?

E→ Identify the dependent and independent variables: babysitting fees and time.

F→ Identify the dependent and independent variables: apples sold by the pound.

G→ Identify the dependent and independent variables: total pay and hours worked.

H→ Identify the dependent and independent variables: cost and tickets purchased.

I→ Given $b(t) = 8t - 2$, evaluate $b(0)$.

J→ Given $f(x) = 3x - 2$, evaluate $f(-5)$.

K→ Given $g(x) = x + 12$, evaluate $g(-2)$.

L→ Given $f(x) = 4x + 7$, evaluate $f(-8)$.

M→ Make a mapping diagram with five coordinate pairs that are not a function.

N→ Give an example of a relationship that is not a function.

O→ Give an example of a relationship that is a function.

P→ Explain why $y = |x|$ is a function and give several examples to support it.

Q→ Given $f(x) = 3x^2 - 2$, evaluate $f(-2)$.

R→ Given $g(x) = x^3 + 12$, evaluate $g(2)$.

S→ Given $h(t) = -t^2 + 1$, evaluate $b(2)$.

T→ Given $b(x) = 2x^2 + 4$, evaluate $b(-4)$.

U→ Given $f(x) = 3x - 2$ and $g(x) = x + 9$, find $f(g(0))$.

V→ Given $f(x) = 2x + 1$ and $g(x) = 3x + 3$, find $g(f(-1))$.

W→ Given $g(x) = 5x + 1$ and $h(x) = 6x - 8$, find $b(g(1))$.

X→ Given $g(x) = 4x + 7$ and $f(x) = x - 5$, find $f(g(-2))$.

Introduction to Functions and Relationships

Lesson 3: Evaluating Functions and Function Notation

Exit Slip

1. The function $f(t) = 50 + 25t$ can be used to model the amount saved $f(t)$ in dollars for any given time t in weeks.

   a. Explain what $f(0)$ represents.

   b. Explain how much money will be in the account after 8 weeks.

   c. Write your answer to "b" using function notation.

2. Given the functions $f(x) = -2x + 2$ and $g(x) = 3x - 1$, describe how to simplify $f(x) - g(x)$ and then simplify it.

Lesson 4: Wrap Up and Assessment
Think Dots

Directions: In your groups, each student will take turns rolling the die. Answer the corresponding question on the record sheet. You may discuss your answers with your group members before recording your answer. You must answer all of the questions.

Take the following data and put it into a table, make a mapping diagram, and sketch a graph. Then identify the domain and range. $(-2,5),(-4,7),(4,7),(2,5)$	Fill in the missing values in the table to represent a function. <table><tr><td>x</td><td>y</td></tr><tr><td>0</td><td>20</td></tr><tr><td>−1</td><td>2</td></tr><tr><td></td><td>1</td></tr><tr><td>4</td><td>15</td></tr><tr><td>7</td><td></td></tr><tr><td>8</td><td>14</td></tr></table>	Show how the vertical line test works for a graph that represents a function and a graph that does not represent a function.
The price of the candy is \$1.89 per pound. a. Determine the dependent and independent variables. b. Write a function to model the situation using function notation: $f(x)=$	Evaluate the function $f(x)=2x+2$ for the following. a. $f(0)$ b. $f(-2)$ c. x when $f(x)=-14$ d. Explain how to use a graphing calculator to find these answers.	Use the functions $f(x)=2x-1$ and $g(x)=4x+5$ to simplify the following. a. $f(x)+g(x)$ b. $f(x)-g(x)$ c. $2f(x)+4g(x)$

Introduction to Functions and Relationships

Lesson 4: Wrap Up and Assessment
Think Dots

Directions: In your groups, each student will take turns rolling the die. Answer the corresponding question on the record sheet. You may discuss your answers with your group members before recording your answer. You must answer all of the questions.

<table>
<tr>
<td></td>
<td></td>
<td></td>
</tr>
<tr>
<td>Take the following data and put it into a table, make a mapping diagram, and sketch a graph. Then identify the domain and range.

$$(-2,5),(-4,7),(4,7),(2,5)$$</td>
<td>Create a set of data that represents a function and a set of data that does not represent a function. Display the data as ordered pairs, in a table, or with a mapping diagram. You must have at least five relationships in your set.</td>
<td>Describe how to use the vertical line test to check if a relationship is a function. Show how the vertical line test works for a graph that represents a function and a graph that does not represent a function.</td>
</tr>
<tr>
<td></td>
<td></td>
<td></td>
</tr>
<tr>
<td>The amount of rain in the rain gauge increases as the rain falls at a rate of 0.75 inches per hour.
a: Determine the dependent and independent variables.
b. Write a function to model the situation using function notation.</td>
<td>Evaluate the function $f(x) = -2x - 4$ for the following.
a. $f(0)$
b. $f(-2)$
c. x when $f(x) = -14$
d. Create your own evaluation problem like the one above and then provide the answers.</td>
<td>Use the functions $f(x) = 3x - 3$ and $g(x) = -2x + 7$ to simplify the following.
a. $f(x) + g(x)$
b. $f(x) - g(x)$
c. $2f(x) + 4g(x)$</td>
</tr>
</table>

★

Lesson 4: Wrap Up and Assessment
Think Dots

Directions: In your groups, each student will take turns rolling the die. Answer the corresponding question on the record sheet. You may discuss your answers with your group members before recording your answer. You must answer all of the questions.

Identify a realistic domain and range for the following situation: An oven set at 350 degrees is turned off after the brownies are done cooking and removed from the oven.	Create a set of data that represents a function and a set of data that does not represent a function. Display the data as ordered pairs, in a table, or with a mapping diagram. You must have at least five relationships in your set.	Explain why the vertical line test works to check if a relationship is a function. Show how the vertical line test works for a graph that represents a function and a graph that does not represent a function.
Create an example of an everyday function that has not been discussed in class. a. Identify the dependent and independent variables. b. Write a function to model the situation using function notation.	Evaluate the function $f(x) = -2x - 4$ for the following. a. $f(0)$ b. $f(-2)$ c. x when $f(x) = -14$ d. Create your own evaluation problem like the one above and then provide the answers.	Use the functions $f(x) = 3x - 3$ and $g(x) = -2x + 7$ to simplify the following. a. $f(g(5))$ b. $g(f(-1))$ c. $2f(x) + 4g(x)$

Introduction to Functions and Relationships

Lesson 4: Wrap Up and Assessment

Think Dots Record Sheet

Circle worksheet chosen:

Introduction to Functions and Relationships

Lesson 4: Wrap Up and Assessment
TriMind Activity

Learning Goal: Understand that everyday situations are represented by functions and relationships.

Directions: Select one of the following activities to complete.

Creative	Analytical	Practical
Create a board game that students can use to test their knowledge of functions and evaluating functions. Your final product will be a creative board game that four players can play. The game should be on poster board or cardboard.	Explain why the vertical line test works to test if a set of relations is a function. Use an example to support your explanation. Your final product will be a poster board with an explanation and accompanying graphs demonstrating how the test works.	Choose an everyday example of a function that has not been discussed in class. Determine a reasonable domain and range and graph the function. Your final product can be presented on a poster board, as a newspaper article, or in any other preapproved method.

Introduction to Functions and Relationships

Lesson 4: Wrap Up and Assessment

TriMind Activity Rubric

Creative	Outstanding (5 points)	Good (3 points)	Poor (0 points)	Points Earned
Creativity	The final product was very creative and original.	The final product showed some creativity.	The final product lacked creativity or was not original.	
Demonstrates understanding of functions	The game demonstrated a strong understanding of functions.	The game contained an error or demonstrated a slight misunderstanding.	The game contained two or more errors or demonstrated a lack of understanding.	
Neatness	The game board was very neat.	The product could have been neater but is readable.	The product was hard to read.	
			Total: _____/15= _____%	

Comments:

Analytical	Outstanding (5 points)	Good (3 points)	Poor (0 points)	Points Earned
Complete explanation	The final product had a complete explanation.	The explanation was missing one element.	The explanation was missing two or more elements.	
Example included	The example supported the explanation well.		The example did not support the explanation.	
Neatness	The product was very neat.	The product could have been neater but was readable.	The product was hard to read.	
			Total: _____/15= _____%	

Comments:

Name: _____ Hour/Block: _____ Date: _____

Practical	Outstanding (5 points)	Good (3 points)	Poor (0 points)	Points Earned
Creativity of everyday application	The application was original.	The application was similar to those discussed in class.	The application was not original.	
Realistic domain and range	The domain and range were realistic.	The domain or range was stated but it was misleading.	The domain or range was not realistic.	
Neatness	The product was very neat.	The product could have been neater but was readable.	The product was hard to read.	
			Total: _____/15= _____%	

Comments:

Lesson 4: WRap Up and Assessment

Exit Slip

1. Give an everyday example of a function.

2. Discuss three significant things that you learned about functions in this unit.

3. What about functions still confuses you?

Unit 2

Systems of Linear Equations

This unit focuses on solving systems of linear equations. Most students will have had experience with solving one-variable equations and graphing linear equations with two variables. In this unit, students will be required to assemble their previous knowledge of solving equations and graphing lines in order to solve systems of linear equations.

This unit begins with pretest and then a whole-class activity that reminds students how to test solutions to equations and introduces them to the idea that two equations can have a common solution. Many of the activities will offer the students an opportunity to choose learning activities according to their learning style and personal interests.

What Do We Want Students to Know?

Common Core State Standards Addressed:	Big Ideas
• 8.EE.7 • 8.EE.8 • A.REI.6 • A.CED.2	• It's important to have a reliable method for solving systems of linear equations. • In systems of equations, there can be one, zero, or infinitely many solutions. • Systems of equations can be used to make solving word problems easier.
	Essential Questions • How do I know which method to use to solve a system? • After solving a system, how do I know if there is/are one, zero, or infinitely many solutions? • When should I use a system to solve word problems?

39

Critical Vocabulary

Solutions
Dependent
Consistent
System of equations

Independent
Inconsistent
Solving using substitution
Solving using elimination

Solving by graphing
Break-even point

Unit Objectives

As a result of this unit, students will know:
➤ how to solve a system of linear equations using graphing,
➤ how to solve a system of linear equations using substitution,
➤ how to solve a system of linear equations using elimination, and
➤ how to identify variables and write a system.

As a result of this unit, students will understand that:
➤ any method can be used to solve a system of equations;
➤ systems of linear equations can model real-life situations;
➤ systems with two variables and two equations can have one, zero, or an infinite number of solutions; and
➤ systems can also include three variables and three equations.

As a result of this unit, students will be able to:
➤ determine the method of solving that will be most efficient,
➤ rewrite equations to make solving a system easier,
➤ write a system of linear equations story problem for a given situation,
➤ solve systems of linear equations story problems, and
➤ determine the number of solutions for the system.

Launch Scenarios

Farmer Bob's Barnyard Problem

Have students read the story problem and then discuss possible strategies for solving it with the person sitting next to them.

> Farmer Bob has a combined total of 40 llamas and chickens in his barnyard. There are a total of 106 legs between the llamas and chickens. Assume that all of the llamas have four legs and all

of the chickens have two legs. How many of each animal are there in the barnyard?

Ask for strategies to solve this story problem. Discuss each strategy. Don't give the final answer, but have students save their guesses. (Lesson 2)

Break-Even Problem

Students will learn about a real-life application of solving a system using graphing. Only the first quadrant will be graphed in this problem, so this is a good opportunity to discuss why it does not make sense to graph negative x- and y-values.

Introduce the situation with the opening questions:

➤ How does a business set its prices for the products and services it sells?
➤ What does it mean for a business to make a profit?
➤ How does a business lose money?

Read the story problem below to the class and then have students talk with their neighbor to determine how best to solve this problem:

> One of the most important things a business does is establish the prices for the products and services it sells to its customers. If a business's total cost is greater than its revenue, then the business loses money. If the revenue is greater than the total cost, then the business makes a profit. The point at which cost and revenue are equal is called the break-even point.

> Kathy wants to organize a trip to Niagara Falls for her family reunion. She found that the fixed costs (i.e., for travel, fees for exhibits and guided tours) for the trip are $5,000, and the additional cost per person (i.e., for meals and accommodations) is $250. She plans to charge $750 per person for the trip. How many people are needed for her to break even? (Lesson 2)

Unit Overview: Systems of Linear Equations

(Assumes a 50–60-minute class period)

Lesson	Whole-Class/Small-Group Discussion and Activities	Individualized Learning Activities
Preassessment (.5 period)	Give the preassessment at least several days prior to beginning the unit (20 minutes)	
Lesson 1: Introduction (2 periods)	Share results of the preassessment (20 minutes) Introduce Farmer Bob's Barnyard Problem (10 minutes)	
		Learning Targets and Study Guide (15 minutes) Vocabulary Choice Board (15 minutes)
	Checkerboard Activity (30 minutes) Student Coordinate Grid Activity (25 minutes)	
Lesson 2: Solving Using Graphing (1.5 periods)	Introduce Break-Even Problem (20 minutes) Solve Farmer Bob's Barnyard Problem (10 minutes) Hold a class discussion on solving using graphing (20 minutes)	
		Tic-Tac-Toe Board (30 minutes) Graphing Systems (10–15 minutes) Exit Slip (10–15 minutes)

Lesson	Whole-Class/Small-Group Discussion and Activities	Individualized Learning Activities
Lesson 3: **Solving Using Substitution** (2 periods)	Hold a class discussion on solving using substitution (20 minutes)	
		Agenda (30 minutes)
		Exit Slip (5–10 minutes)
Lesson 4: **Solving Using Elimination** (1 period)	Hold a class discussion on solving using elimination (20 minutes)	
		Tic-Tac-Toe Board (30 minutes)
		Exit Slip (5–10 minutes)
Lesson 5: **Mixture Problems** (1 period)	Hold a class discussion on solving mixture problems (20 minutes)	
		Agenda (40 minutes)
		Exit Slip (5–10 minutes)
Lesson 6: **Wrap Up and Assessment** (2–4 periods)	Collect learning targets and study guide handouts and vocabulary choice board projects	
		Real-Life Applications (50 minutes)
		Think Dots (40 minutes)
		RAFT Activity (40 minutes)
		Small Business Plan (50 minutes)
		Exit Slip (5–10 minutes)

Systems of Linear Equations
Preassessment

Directions: Solve the following problems to the best of your ability, and then rate your confidence in your answer in the space provided. Skip questions you do not know. This will not be graded.

1. The coordinate pair $(2,-4)$ is a solution to the system $\begin{cases} y = 2x + 2 \\ y = -3x + 2 \end{cases}$

   _____ I'm sure

   _____ I'm not sure

2. Solve the following system using a graphing calculator: $\begin{cases} y = -2x + 1 \\ -y = 3x + 2 \end{cases}$

   _____ I'm sure

   _____ I'm not sure

3. Solve the following system using substitution: $\begin{cases} y = -4x + 5 \\ y + 3x = 2 \end{cases}$

   _____ I'm sure

   _____ I'm not sure

4. Solve the following system using elimination: $\begin{cases} y = -2x + 1 \\ y + 3x = 2 \end{cases}$

   _____ I'm sure

   _____ I'm not sure

5. Write a system of equations to model the following situation and then solve it.

   The drama club is selling tickets to its latest production. Adult ticket prices are $5.50 and a ticket for a child is $3. The club sold a total of 443 tickets for its weekend shows and collected a total of $1,566.50. How many of each type of ticket was sold?

   _____ I'm sure

   _____ I'm not sure

6. Write a system of equations to model the following mixture problem and then solve it.

   In your chemistry class, you are given a bottle of 5% solution and a bottle of 2% solution. You need 60 ml of a 3% solution for an experiment. How much of each solution do you need to mix together to get the desired concentration?

   _____ I'm sure

   _____ I'm not sure

Name: _____ Hour/Block: _____ Date: _____

Systems of Linear Equations
Learning Targets and Study Guide

At the end of this unit, every student should be able to say:

Learning Target	Explain and Give an Example
I can determine if an ordered pair is a solution to a system of linear equations. Initial here when mastered _____	
I can solve systems of equations using graphing. Initial here when mastered _____	
I can solve systems of equations using substitution. Initial here when mastered _____	
I can solve systems of equations using elimination. Initial here when mastered _____	
I can solve mixture problems. Initial here when mastered _____	
I can solve real-life story problems using systems of equations. Initial here when mastered _____	

Systems of Linear Equations

Systems of Linear Equations
Vocabulary Choice Board

Directions: Using the critical vocabulary for the unit, choose one of the following methods to demonstrate your knowledge of the mathematical terms.

LINCS Cards	Graphic Organizer	News Broadcast
Prepare a LINCS card for each vocabulary word to demonstrate your understanding of the vocabulary words. *(logical-mathematical)*	Construct a graphic organizer of your choice to define each word and demonstrate your understanding of the vocabulary words and how they fit into the current unit. *(verbal/language-mathematical)*	Write a news broadcast that defines all of the vocabulary words, and provide stories that demonstrate your understanding of the vocabulary words. Turn in your news broadcast, and then present it to the class. *(intrapersonal-language)*
Nature Survival	**Creative Story**	**Poster Board**
Prepare a nature survival guide about how each of the vocabulary words would be used if you were stranded on a deserted island. The guide must demonstrate your understanding of the vocabulary words. *(natural-language)*	Write a creative story with color illustrations that demonstrates your understanding of the vocabulary words. *(verbal-language)*	Prepare an 11" x 14" poster board that contains the definitions of the vocabulary words and includes one real-life example for each word that demonstrates your understanding of the definition. *(language-spatial)*
Cartoon	**Rap or Song**	**Acting Out**
Prepare a cartoon strip with at least five frames that defines each word and is illustrated with appropriate pictures to demonstrate your understanding of the vocabulary words. *(visual-spatial)*	Write a rap or song containing all of the vocabulary words and their correct definitions. Perform your rap or song for the class. *(musical-language)*	Write a script and then act out the words with their correct definitions. Use props if appropriate. *(kinesthetic-language)*

Critical Vocabulary

Solutions
Dependent
Consistent
System of equations

Independent
Inconsistent
Solving using substitution
Solving using elimination

Solving by graphing
Break-even point

Systems of Linear Equations
Vocabulary Choice Board Rubric

Grading scale:	3 Well done	2 Close to requirements		1 Needs more work

Choice	Definitions	Illustrations and/ or creativity	Neatness	Other
LINCS cards	Definitions demonstrate knowledge of vocabulary. Score _____	The illustrations are appropriate and creative. Score _____	The cards are easy to read. Score _____	Semantic connections between words are demonstrated. Score _____
Graphic organizer	Definitions demonstrate knowledge of vocabulary. Score _____	The graphic organizer creatively displays the definitions. Score _____	The graphic organizer is easy to read. Score _____	The graphic organizer choice is appropriate for this assignment. Score _____
News broadcast	Definitions demonstrate knowledge of vocabulary. Score _____	The news broadcast is creative and holds the attention of the audience. Score _____	The news broadcast is easy to read. Score _____	Good presentation skills are demonstrated. Score _____
Nature survival guide	Definitions demonstrate knowledge of vocabulary. Score _____	The guide contains illustrations that are appropriate. Score _____	The guide is easy to read. Score _____	The final product resembles a survival guide. Score _____
Creative story	Definitions demonstrate knowledge of vocabulary. Score _____	The story contains illustrations that are appropriate. Score _____	The story is easy to read. Score _____	The story is interesting. Score _____
Poster board	Definitions demonstrate knowledge of vocabulary. Score _____	The poster board includes illustrations that are appropriate. Score _____	The poster board is easy to read. Score _____	Real-life examples are included for each vocabulary term. Score _____
Cartoon	Definitions demonstrate knowledge of vocabulary. Score _____	The cartoon includes illustrations that are appropriate. Score _____	The cartoon is easy to read. Score _____	The cartoon contains a minimum of five frames. Score _____
Rap or song	Definitions demonstrate knowledge of vocabulary. Score _____	The rap or song is creative. Score _____	The lyrics are easy to read. Score _____	Good performance skills are demonstrated. Score _____
Acting out	Definitions demonstrate knowledge of vocabulary. Score _____	The script is creative. Score _____	The script is easy to read. Score _____	Good performance skills are demonstrated. Score _____

Total: _____ /12= _____%

Comments:

Lesson 1: Introduction

Checkerboard Small-Group Activity Lesson Plan

Purpose: Students will work in small groups (in pairs or groups of three) to put together the checkerboard puzzle that reviews identifying solutions to linear equations and identifying the slope and y-intercept from an equation.

Prerequisite Knowledge	Materials Needed
Students should know how to: • solve equations for y, • check solutions to equations, • find the slope and y-intercept from equations in the form $y = mx + b$, and • find slopes of vertical and horizontal lines.	• One copy of the checkerboard for each small group of students, cut into squares and mixed up • Scrap paper for solving equations

Discussion Questions (To be asked before the lesson.)
- ➤ How do you determine whether an ordered pair is a solution to an equation?
- ➤ How do you rewrite an equation that is not in the form $y = ?$
- ➤ How do you find slope and the y-intercept from an equation written in the form $y = mx + b$?
- ➤ What is the slope of a horizontal line? Give an example of an equation for a horizontal line.
- ➤ What is the slope of a vertical line? Give an example of an equation for a vertical line.

Lesson
- ➤ Teachers may want to use random grouping. Each group should receive a packet of the puzzle pieces that are in no particular order.
- ➤ Students are asked to put together the checkerboard by matching pieces that have equations with a possible solution or by matching an equation with the slope or y-intercept . Their finished puzzle is the message "Math Does Rock Says Me 2 U" (with the help of some text messaging shortcuts).
- ➤ Students may notice that there are puzzle pieces that are blank on one side. They may figure out that these pieces belong on the outer edge of the puzzle.

Discussion Questions (To be asked after the lesson.)
- ➤ What message is revealed when the puzzle is completed?
- ➤ How do you determine whether an ordered pair is a solution to an equation? How many solutions are there to a linear equation?
- ➤ Can all equations be rewritten in the form of $y = ?$ Why would we want to write equations in this form?
- ➤ How do you determine the slope and the y-intercept from an equation?

$3y = -4x + 12$ **M** (0,3) $y = 2x + 3$ $2x + 3y = 6$	$3y = 9x + 11$ $y = 2x + 3$ **A** $2y = 8x$ (1,7)	$y = 2x + 12$ $m = 4$ **T** $m = $ zero	$3y = -2x + 12$ $2y = 3x + 1$ (1,2) **H** $m = \dfrac{1}{3}$
(3,0) **D** $m = \dfrac{3}{2}$ $y = -12$	$y = 2x + 5$ $2y = 3x$ **O** $2x + 4y = 0$	$y = -12$ $m = 3$ **E** (3,−5) and (6,4) $m = 3$	$3y = x + 2$ $x = 2$ $m = $ undefined **S** $2y = 4x$
$b = -12$ **R** $m = $ zero $y = -2$ $2x + 3y = 6$	(−2,1) **O** $m = -12$	$y = 3x + 5$ $2y = 8x + 2$ $b = 1$ **C** $3y = 6x$	(1,2) $m = \dfrac{1}{5}$ **K** $5y = x + 2$ $2y = 6x$
$b = 2$ **S** $m = \dfrac{1}{3}$ $-3y = -2x + 15$	$y = -12x + 5$ $12y = 4x$ **A** $x = 2$	$m = 2$ $3x + 2 = y$ **Y** (1,5) (2,1)	$b = 0$ $y = -4x + 5$ **S** (0,5) $-2y = 4x + 4$
$b = -5$ **M** $m = -2$ $y = 6x - 1$	$m = $ undefined $2y = -4x + 3$ **E** $y = 6x + 1$	$y = -2x + 5$ $m = 2$ **2** $2y = 4x + 1$ $2y = 16x + 15$	$m = -2$ $y = -4x + 5$ **U** $b = 5$ $2y = -6x - 1$

Lesson 1: Introduction

Student Coordinate Grid Whole-Class Activity Lesson Plan

Purpose: Students will model a system of linear equations by using the classroom as a coordinate grid.

Prerequisite Knowledge	Materials Needed
Students should know how to: • solve equations, • check solutions to equations, • graph equations on calculators, and • find coordinates on a coordinate grid.	• One copy of the student worksheet for each student • Note cards with coordinate pairs $(0, 0)$ through $(5, 4)$ • Overhead projector with 6 x 5 grid matching the setup of the desks

Discussion Questions (To be asked before the lesson.)
> ➤ Can two different linear equations have one solution in common?
> ➤ What does this solution mean?

Lesson
> ➤ Write the following equations on the board, and label them for future reference.
>
> Equation A $x + y = 5$
>
> Equation B $x - y = -1$
>
> ➤ Arrange the classroom desks into a coordinate grid. Each student will represent one of the ordered pairs. Below is an example of a class with 30 students. This activity could also be done outside with a grid drawn with sidewalk chalk.

$(0, 4)$	$(1, 4)$	$(2, 4)$	$(3, 4)$	$(4, 4)$	$(5, 4)$
$(0, 3)$	$(1, 3)$	$(2, 3)$	$(3, 3)$	$(4, 3)$	$(5, 3)$
$(0, 2)$	$(1, 2)$	$(2, 2)$	$(3, 2)$	$(4, 2)$	$(5, 2)$
$(0, 1)$	$(1, 1)$	$(2, 1)$	$(3, 1)$	$(4, 1)$	$(5, 1)$
$(0, 0)$	$(1, 0)$	$(2, 0)$	$(3, 0)$	$(4, 0)$	$(5, 0)$

> ➤ Have students substitute the x-value and y-value of their coordinate pair into the first equation. Instruct students to stand up if their coordinate pair represents a solution to the equation.
> ➤ Ask students what they notice about the students who are standing. Instruct the students to lightly shade the solutions (the spaces where students are standing) on their worksheet.
> ➤ Have the students sit down and then repeat the previous two steps with the second equation.

Discussion Questions (To be asked after the lesson.)
> ➤ Which ordered pair was a solution for both equations?
> ➤ How is this solution reflected using the human coordinate grid?
> ➤ How is this solution reflected on your table?

Lesson 1: Introduction
Student Coordinate Grid

Equation A $x + y = 5$

Equation B $x - y = -1$

Discussion Questions
- ➤ Can two different linear equations have one solution in common?
- ➤ What does this solution mean?

Directions
- ➤ Each student has been assigned a coordinate pair. Write your coordinate pair in the space provided.   (,   )
- ➤ Check to see if your coordinate pair is a solution to Equation A. Show your work. If it is a solution, you will be asked to stand. If it is not, you will remain seated.
- ➤ Lightly shade the coordinate pairs in the table below that represent the students who are standing after testing their coordinate pair with Equation A.

(0, 4)	(1, 4)	(2, 4)	(3, 4)	(4, 4)	(5, 4)
(0, 3)	(1, 3)	(2, 3)	(3, 3)	(4, 3)	(5, 3)
(0, 2)	(1, 2)	(2, 2)	(3, 2)	(4, 2)	(5, 2)
(0, 1)	(1, 1)	(2, 1)	(3, 1)	(4, 1)	(5, 1)
(0, 0)	(1, 0)	(2, 0)	(3, 0)	(4, 0)	(5, 0)

- ➤ Repeat the second step with your coordinate pair and Equation B. Show your work. Again, you will be asked to stand if your coordinate pair is a solution, and you will remain seated if it is not.
- ➤ Lightly shade the coordinate pairs in the same table that represent the students who are standing after testing their coordinate pair with Equation B.

Discussion Questions
- ➤ Which ordered pair was a solution for both equations?

- ➤ How is this solution reflected using the human coordinate grid?

- ➤ How is this solution reflected on your table above?

- ➤ What does this solution mean?

Systems of Linear Equations

Lesson 2: Solving Systems Using Graphing
Tic-Tac-Toe Board

Directions: Choose three problems in a row, in a column, or on a diagonal. Solve the problems and show your work on a separate sheet of paper. Circle the problems you choose on this page and then reference them on your work paper with the number from the square.

1	2	3
Write the steps to follow when using a graphing calculator to solve the following system. $$\begin{cases} -y = -2x \\ x - y = -6 \end{cases}$$	Write a story problem similar to the Farmer Bob's Barnyard Problem and show how to solve it using graphing.	Write a system of linear equations that has a solution at the origin. Draw the graph and label the solution.
4	**5**	**6**
Solve the system $$\begin{cases} -2y - 2x = 4 \\ 2x - y = -4 \end{cases}$$ using graphing. Draw the graph and label the solution.	Write a system of linear equations that includes a vertical line and a solution of $(1, 2)$. Draw the graph and label the solution.	Write two systems of equations: one that has no solutions and one that has an infinite number of solutions. Draw the graphs of these two situations.
7	**8**	**9**
Write a break-even story problem for a business that you will own when you finish high school. Then show how to solve it using graphing.	Write a system of linear equations that has a solution of $(3, 2)$. Draw the graph and label the solution.	Write a system of linear equations that includes a horizontal line and a solution of $(1, 2)$. Draw the graph and label the solution.

Lesson 2: Solving Using Graphing

Graphing Systems

Directions: Solve each system by graphing. You must graph each line by hand on the grids, and then check your answer using a graphing calculator. Label the solution on the grid.

1. $\begin{cases} y = 2x - 1 \\ y = -2x + 1 \end{cases}$

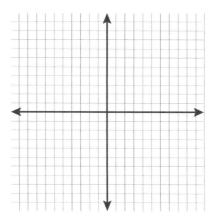

2. $\begin{cases} y = 3x - 3 \\ y = -3x - 6 \end{cases}$

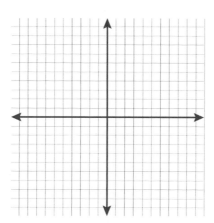

3. $\begin{cases} y = \dfrac{4}{3}x + 3 \\ y = -\dfrac{2}{3}x - 3 \end{cases}$

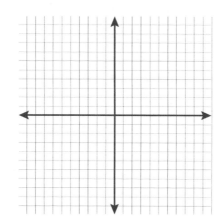

Systems of Linear Equations

4. $\begin{cases} y = 4x - 1 \\ y = -x + 4 \end{cases}$

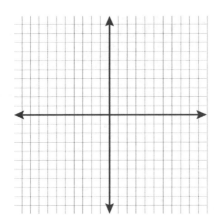

5. $\begin{cases} y = \dfrac{5}{4}x - 2 \\ y = \dfrac{5}{4}x + 1 \end{cases}$

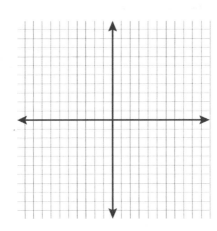

6. $\begin{cases} -x + 2y = 6 \\ x + 4y = 24 \end{cases}$

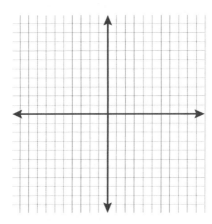

Lesson 2: Solving Using Graphing
Graphing Systems

Directions: Solve each system by graphing. You must graph each line by hand on the grids, and then check your answer using a graphing calculator. Label the solution on the grid.

1. $\begin{cases} y = 2x - 1 \\ y = -2x + 1 \end{cases}$

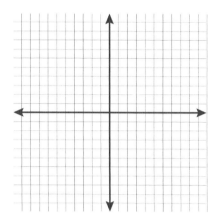

2. $\begin{cases} y = \dfrac{4}{3}x + 3 \\ y = -\dfrac{2}{3}x - 3 \end{cases}$

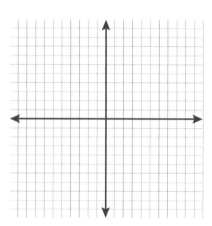

3. $\begin{cases} y = 6x - 5 \\ -2x + y = -2 \end{cases}$

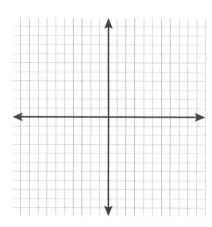

Systems of Linear Equations

4. $\begin{cases} 2y + 4x = 8 \\ 12x = -6y + 24 \end{cases}$

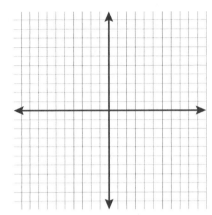

5. $\begin{cases} y = \dfrac{5}{4}x - 2 \\ y = \dfrac{5}{4}x + 1 \end{cases}$

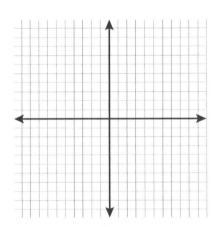

6. $\begin{cases} -x + 2y = 6 \\ x + 4y = 24 \end{cases}$

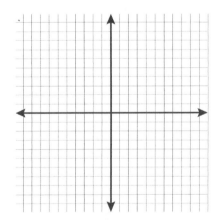

Systems of Linear Equations

Lesson 2: Solving Using Graphing
Graphing Systems

Directions: Solve each system by graphing. You must graph each line by hand on the grids, and then check your answer using a graphing calculator. Label the solution on the grid.

1. $\begin{cases} y = 2x - 1 \\ y = -2x - 1 \end{cases}$

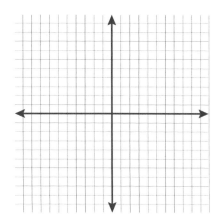

2. $\begin{cases} y = \dfrac{4}{3}x + 3 \\ y = -\dfrac{2}{3}x - 3 \end{cases}$

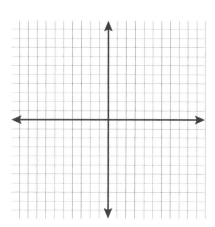

3. $\begin{cases} y = 6x - 5 \\ -2x + y = -2 \end{cases}$

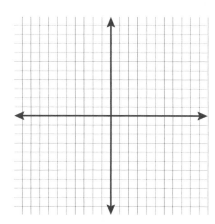

Systems of Linear Equations

4. $\begin{cases} 2y + 4x = 8 \\ 12x = -6y + 24 \end{cases}$

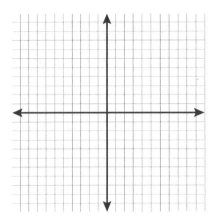

5. $\begin{cases} y + \dfrac{1}{2}x = -2 \\ 3y - 6x = -15 \end{cases}$

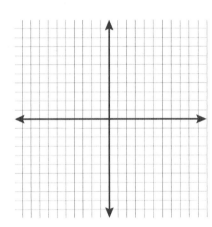

6. $\begin{cases} -4x = -2y + 3 \\ x = -2y - 1 \end{cases}$

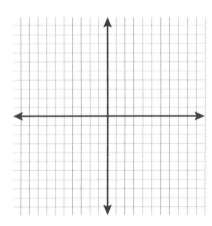

Lesson 2: Solving Using Graphing
Exit Slip

1. Explain when to use graphing to solve systems of linear equations and how to find the solution.

2. Give an example of a system that could be easily solved using graphing.

Lesson 3: Solving Using Substitution
Agenda

Imperatives (You must do all three of these.)

1. Write systems of equations that meet the following requirements.
   a. consistent
   b. inconsistent
   c. dependent
   d. independent

2. Solve the following systems using substitution. Check your solutions by graphing or by using the original equations.

   a. $\begin{cases} y = 3x + 9 \\ y = -5x - 7 \end{cases}$

   b. $\begin{cases} 4x = y - 1 \\ 6x - 2y = -3 \end{cases}$

   c. $\begin{cases} \dfrac{1}{2}x + \dfrac{1}{3}y = 6 \\ x - y = 2 \end{cases}$

3. Solve the following story problem using substitution. Check your answer to be sure it is reasonable.

   Your family is trying to decide between two cable companies. Satellite Systems charges $49 to set up the service and $29 per month for basic cable. The other company, Communications Cable, charges $25 to set up the service and $37 per month for basic cable.

   a. In how many months will both of the services cost the same?
   b. What will that cost be?
   c. Which system should you choose if your family will only need the service for 6 months?

Negotiables (You must do at least two of these.)

1. A class of 342 students went on a field trip using both buses and vans. Each bus holds 53 students and a van can hold 8 students. A total of 9 vehicles were used. Write and solve a system of equations to determine how many of each type of vehicle was used.

2. Mitch's Skate Shop sells skateboards and wakeboards. Wakeboards are sold for $49, and skateboards are sold for $129. The revenue from these skateboards totaled $6,965, and a total of 85 skateboards and wakeboards were sold. Write and solve a system of equations to determine how many of each product was sold.

3. A dance studio offers classes in jazz and hip hop. There are currently 30 people in the jazz class, and the class size has been increasing by 2 people each week. The hip hop class currently has 10 people, but it has been increasing by 3 people per week. Write and solve a system of equations to determine how many weeks it will take for the number of people in each class be the same.

Systems of Linear Equations

Options (You may choose to do these problems.)

A magic square is a table with the same number of columns and rows. Each row, column, and diagonal in a magic square adds to the same number. Use algebra to determine the value of the variables in the magic squares. An example has been provided for you to follow.

Example:

−4	−6	B
6	−2	−10
A	2	C

> Find a row or column that has no variables and add the numbers. For this example, the row is highlighted. The total is (−6).
> Write and solve an equation for the first column and make it equal to the sum you found.

$$-4 + 6 + A = -6$$
$$2 + A = -6$$
$$A = -8$$

> Do the same with the first row to solve for B.

1. Find the missing values in the magic square.

2	B	−2
A	−1	C
0	1	−4

2. Find the missing values in the magic square.

A	11	8
C	B	−7
E	D	14

Lesson 3: Solving Using Substitution
Exit Slip

1. Explain how to solve systems of linear equations using substitution. Use an example system in your explanation.

2. Give two more examples of systems that could be easily solved using substitution.

Lesson 4: Solving Using Elimination
Tic-Tac-Toe Board

Directions: Choose three problems in a row, in a column, or on a diagonal. Solve the problems and show your work on a separate sheet of paper. Circle the problems you choose on this page and then reference them on your work paper with the number from the square.

1	**2**	**3**
Solve: $$\begin{cases} 4x + 3y = 16 \\ 2x - 3y = 8 \end{cases}$$	The Pinz bowling alley charges $4 for shoe rental plus $2 per game. Bowl-a-Mania charges $2.50 per game plus $2.00 for shoe rental. How many games would you have to bowl for the cost to be the same at both? What would be the cost?	Katie spent $465 on DVDs and CDs. The CDs cost $15 each, and the DVDs cost $19 each. A total of 27 CDs and DVDs were purchased. How many of each did she buy?
4	**5**	**6**
A jar contains dimes and quarters. There are a total of 28 coins. The total value of the coins is $5.05. How many of each type of coin is in the jar?	Solve: $$\begin{cases} 3x + 2y = 8 \\ 2y = 12 - 5x \end{cases}$$	Solve: $$\begin{cases} -x - 5y - 5z = 2 \\ 4x - 5y + 4z = 19 \\ x + 5y - z = -20 \end{cases}$$
7	**8**	**9**
Solve: $$\begin{cases} 5y - 20 = -4x \\ y = -\dfrac{5}{4}x + 4 \end{cases}$$	Erin worked twice as many hours as her friend Katey plus one more hour. Three times the number of hours that Katey worked plus the amount of hours that Erin worked equals six. How many hours did Erin and Katey work?	The sum of two numbers is 66. The first number is 24 less than twice the second number. Find the two numbers.

Systems of Linear Equations

Lesson 4: Solving Using Elimination
Exit Slip

1. Explain how to solve systems of linear equations using elimination. Use an example system in your explanation if you need to.

2. Give two more examples of systems that could be easily solved using elimination.

Systems of Linear Equations

Lesson 5: Mixture Problems
Agenda

Imperatives (You must do all three of these.)

1. Change the following percents to decimals.

   a. 10% _____   f. 25% _____

   b. 1% _____   g. 0.6% _____

   c. 0.06% _____   h. 100% _____

   d. 135% _____   i. 200% _____

   e. 1000% _____   j. 2.5% _____

2. Write an equation to represent the following questions. Then solve each equation.

   a. What is 50% of 10? _____   e. 10 is what percent of 100? _____

   b. What is 25% of 40? _____   f. 4 is what percent of 50? _____

   c. What is 75% of 100? _____   g. 15 is 30% of what number? _____

   d. What is 1% of 1006? _____   h. 90 is 75% of what number? _____

3. Solve the following mixture problem using substitution or elimination. Check your answer to be sure it is reasonable.

   In your chemistry class, you are given a bottle of 5% acid solution and a bottle of 2% acid solution. You need 60 ml of a 3% solution for an experiment. How much of each solution do you need to mix together to get the desired concentration?

Negotiables (You must do at least two of these.)

1. Seven pounds of a candy mixture containing 30% M&Ms was mixed with 2 pounds of a candy mixture containing 20% M&Ms. What percent of the new mixture is M&Ms? (Round your answer to the nearest whole percentage.)

2. Zach mixed together 4 gallons of a juice concentrate containing 25% pineapple juice with 7 gallons of a juice concentrate containing 38% pineapple juice. What is the pineapple juice concentration of the new mixture? (Round your answer to the nearest whole percentage.)

3. How many mg of a metal containing 40% copper must be combined with 8 mg of pure copper to form an alloy containing 65% copper? (Hint: What does it mean if a metal is pure? Round your answer to the nearest tenth.)

Systems of Linear Equations

Options (You may choose to do these problems.)

1. Adam's Coffee and Nuts produces and sells a 24-oz. bag of mixed nuts that contain 25% cashews. In order to make this mixture, they combine a premium mix that contains 50% cashews with an economy mix that contains only 19% cashews. How much of each do they need to use? (Round your answer to the nearest tenth.)

2. A metallurgist needs to make 16.6 ounces of an alloy containing 40% nickel. If she melts a metal that is 30% nickel with another metal that is 50% nickel, how much of each should she use?

Lesson 5: Mixture Problems
Exit Slip

1. Explain the two equations that are typically used to solve a mixture problem. Use an example if necessary.

2. Give another example of a mixture problem.

Lesson 6: Wrap Up and Assessment
Real-Life Applications

Directions: Attached are eight story problems that require first writing and then solving systems of linear equations. You must choose five of these problems and provide the information outlined in the scoring rubric. Below is a story problem that will model what is expected of you for the five story problems you choose. Each step must be labeled. Neatness is very important. Record all of the following steps as the class solves the story problem.

Step 1: At the high school track meet, 146 hot dogs and hamburgers were sold. A total of $238 was collected. Hamburgers sell for $2 and hot dogs sell for $1. How many of each were sold?

Step 2: 146 items were sold; $2 for burgers; $1 for hot dogs; $238 was collected.

Step 3: h = hot dogs; b = burgers

Step 4: $h + b = 146$; $2b + h = 238$

Step 5: Substitution: Elimination:

$h = 146 - b$ $h + b = 146$

$2b + h = 238$ $h + 2b = 238$

$b = 92$ and $h = 54$

Step 6: Substitution: Elimination:

$h + b = 146$ $2b + h = 238$

$54 + 92 = 146$ $2(92) + 54 = 238$

$146 = 146$ $184 + 54 = 238$

$238 = 238$

Step 7: They sold 92 burgers and 54 hot dogs.

Systems of Linear Equations

Lesson 6: Wrap Up and Assessment
Real-Life Applications

Directions: Choose five of the following story problems to complete.

1. Katie has 25 coins that are dimes and nickels. Together they total $2.00. How many of each type of coin does she have?

2. The sum of two numbers is 90. Their difference is 34. What are the two numbers?

3. The drama club sold 288 tickets to its spring play and collected $1,161. Adult tickets cost $5.00 each and children's tickets cost $3.50. How many of each type of ticket did the drama club sell?

4. In your chemistry class, you are given a bottle of 5% hydrochloric acid solution and a bottle of 2% hydrochloric acid solution. You need 60ml of a 3% solution for an experiment. How much of each solution do you need to mix to get the solution you need?

5. Your school's student council is selling flower bulbs for a school fundraiser. Customers can buy bags of tulip bulbs or bags of daffodil bulbs. Eric sold 10 bags of tulip bulbs and 8 bags of daffodil bulbs and collected $121. Tom sold 5 bags of tulip bulbs and 12 bags of daffodil bulbs and collected $116.50. What is the cost for one bag of each of the types of flower bulbs they are selling?

6. One high-speed Internet provider charges a set-up fee of $180 and $40 per month for service. The other provider charges a set-up fee of $100 and $60 per month for service. In how many months will the cost of both plans be the same?

7. All 330 freshmen at a high school went on a field trip. Some students rode in vans that hold 12 students. Others rode in buses that hold 45 students. How many of each type of vehicle was used if they took a total of 11 vehicles?

8. There are a total of 52 chickens and goats in the barnyard. Assume that each chicken has two legs and each goat has four legs. If there are a total of 128 legs, how many of each animal is in the barnyard?

Systems of Linear Equations

Lesson 6: Wrap Up and Assessment

Real-Life Applications Rubric

Systems of Linear Equations

	Meets Expectations (2 points)	Slightly Below Expectations (1 point)	Below Expectations (0 points)	1	2	3	4	5
1. Rewrite the story problem at the top of the page.	The complete story problem was written.	The story problem was written, but it was not complete.	The story problem was not rewritten.					
2. Summarize the important information.	All important information was summarized.	Some important information was not summarized.	No information was summarized.					
3. Identify the variables (let $x =$ and let $y =$).	Variables were clearly and completely identified with units if needed.	Variables were identified but not completely or units were missing.	Variables were not identified or units were missing.					
4. Write a system of equations.	Two correct equations were written.	Two equations were included but one was not correct.	Equations were not correct.					
5. Solve the system using substitution and elimination.	The system was solved correctly using two methods.	The system was solved correctly but only one method was used.	The system was not solved correctly.					
6. Check your solution with the original equations.	The solution was checked in both equations.	The solution was checked in just one equation.	The solution was not checked or was incorrect.					
7. Explain your answer in a real-world context.	The answer was explained using the context of the problem.	The answer was explained but not completely.	The answer was not explained or did not match the context of the problem.					
Neatness of the entire project (score is multiplied by 5 and added to the total):	Work was neatly done.	Work was difficult to read.	Work could not be read.					
			Total: _____ /80= _____ %					

Comments:

Lesson 6: Wrap Up and Assessment
Think Dots

Directions: In your groups, each student will take turns rolling the die. Answer the corresponding question on the record sheet. You may discuss your answers with your group members before recording your answer. You must answer all of the questions.

Write a system of equations that will result in exactly one solution, a system that has no solutions, and a system that has an infinite number of solutions.	We learned how to solve using graphing, substitution, and elimination. For the following systems, explain the best method to use and why. a. $\begin{cases} 2x + 4y = 10 \\ -2x + 3y = 11 \end{cases}$ b. $\begin{cases} y = 2x - 1 \\ y = -3x + 1 \end{cases}$ c. $\begin{cases} y = x - 1 \\ 2x + 3y = 2 \end{cases}$	Using solving by elimination, what are the possible numbers you could multiply the following system by to eliminate a variable? $\begin{cases} 2x - 3y = 5 \\ -3x + y = 12 \end{cases}$
Solve the following system of equations using substitution. Check your solution. $\begin{cases} y = -2x + 3 \\ y = -4x + 2 \end{cases}$	Create a story problem that can be represented by the following system of equations. $y = 2.50x + 50$ $y = 3.50x + 25$	Explain how to use a graphing calculator to graph and solve the following system. $\begin{cases} y = 5x - 1 \\ 2y = -3x + 1 \end{cases}$

Lesson 6: Wrap Up and Assessment
Think Dots

Directions: In your groups, each student will take turns rolling the die. Answer the corresponding question on the record sheet. You may discuss your answers with your group members before recording your answer. You must answer all of the questions.

Systems of Linear Equations

Write a system of equations that will result in exactly one solution, a system that has no solutions, and a system that has an infinite number of solutions.	Explain how to use a graphing calculator to graph and solve the following system of equations. $$\begin{cases} y = 5x - 1 \\ 5x = 2y - 1 \end{cases}$$	Using solving by elimination, what are the possible numbers you could multiply the following system by to eliminate a variable? $$\begin{cases} 2x - 3y = 5 \\ -3x + 7y = 12 \end{cases}$$
Solve the following system of equations using any method. Check your solution. $$\begin{cases} y = 4x + 2 \\ 4y = 8x - 8 \end{cases}$$	Create a story problem that can be represented by the following system of equations. $$\begin{cases} 1.50x + 2.00y = 225 \\ x + y = 125 \end{cases}$$	At the softball game 128 hot dogs and slices of pizza were sold. Hot dogs cost \$1 and pizza cost \$1.50. A total of \$151 was collected. How many of each were sold? Write a system and solve.

Lesson 6: Wrap Up and Assessment
Think Dots

Directions: In your groups, each student will take turns rolling the die. Answer the corresponding question on the record sheet. You may discuss your answers with your group members before recording your answer. You must answer all of the questions.

Create a story problem that can be represented by the following system of equations. $\begin{cases} .25x + .57y = 90(.35) \\ x + y = 90 \end{cases}$	How many mg of a metal containing 30% silver must be combined with 10 mg of pure silver to form an alloy containing 60% silver? (Hint: what does it mean if a metal is pure?) Round your answer to the nearest tenth.	Using solving by elimination, what are the possible numbers you could multiply the following system by to eliminate a variable? $\begin{cases} 2x = -3y - 12 \\ -3y + 7 = -5x \end{cases}$
Explain the difference between an independent system and a dependent system and give an example of each.	Solve the following system of equations using any method. Check your solution. $\begin{cases} -2y = 4x + 2 \\ 3x = 8 - 4y \end{cases}$	Solve the following system of three equations and three variables. $\begin{cases} x = -4y + 4z + 4 \\ z = 5x - 25 \\ -2x - 5z = 17 \end{cases}$

Systems of Linear Equations

Name: _____ Hour/Block: _____ Date: _____

Lesson 6: Wrap Up and Assessment
Think Dots Record Sheet

Circle worksheet chosen: ★

Lesson 6: Wrap Up and Assessment
RAFT Activity

Congratulations, our unit is nearing completion! You will now have the opportunity to demonstrate your understanding of one key element of the unit by choosing one of the following RAFT activities. If you would like to propose a change to any of the roles, audiences, or formats that are listed, you must have your idea approved before you start your RAFT activity. In addition, any student choosing the "student choice" option at the bottom must complete a Student Choice Proposal Form.

Role	Audience	Format	Topic
Pair of equations	People in the community	Advice column	Everyday applications of systems of linear equations
Graphing calculator	Little brother or sister	Instruction manual	Graphing systems of equations on a calculator and finding the solution
Three different sets of systems	Potential voters	Campaign speech	Which method is best for solving a system
Break-even point	Investment banker	Business plan	How to make a profit
Beaker	Scientists	Recipe	Solving mixture problems
Teacher	Future math students	Brochure	Types of systems (e.g., consistent, dependent)
Solution	System of equations	Instructions	Testing a solution with a set of linear equations
Student choice	Student choice	Student choice	Student choice

Lesson 6: Wrap Up and Assessment

RAFT Activity Rubric

Criteria	Excellent (2 points)	Acceptable (1 point)	Poor (0 points)	Points Earned
Role	The student assumed the appropriate role and wrote from the proper perspective.	The student assumed the role but the perspective was not correct.	The role did not appear related to the assignment.	
Audience	The audience was appropriately addressed.	The audience was addressed but not convincingly.	The intended audience was unclear.	
Format	The student clearly demonstrated an understanding of the format.	The student did not use the format well.	The format was not correct.	
Topic	The student demonstrated an understanding of the topic and communicated it well. The student used sample problems.	The student demonstrated some understanding of the topic. Sample problems were used sparingly.	The student clearly did not understand the topic. The student used very few sample problems.	
Creativity	The student demonstrated original thinking and entertained the audience.	The ideas presented were not original but the audience was entertained.	There was no creativity and the final product was not entertaining to the audience.	
Neatness	The final product was easy to read and understand.	Most of the product was neat and easy to read or understand.	The final product was not easy to read or understand.	
			Total: _____/12= _____%	

Comments:

Systems of Linear Equations

Systems of Linear Equations RAFT Student Choice Proposal Form	Name:
	Teacher's Approval:
	Date:

| **Proposals for Student Choices** |
| Role: |
| Audience: |
| Format: |
| Topic: |

- -Cut Here- -

Systems of Linear Equations RAFT Student Choice Proposal Form	Name:
	Teacher's Approval:
	Date:

| **Proposals for Student Choices** |
| Role: |
| Audience: |
| Format: |
| Topic: |

Lesson 6: Wrap Up and Assessment
Small Business Plan Lesson Plan

Introduction

Many students have been exposed to the idea of a lemonade stand (or other small business), either through personal experience or by playing an online simulation game. The goal of this lesson is to provide students the opportunity to develop a very simplified small business and then determine the amount of product/services that must be sold in order to break even. The break-even problem included at the beginning of this chapter (or a similar problem) should be discussed as a group prior to starting this project. This project could be used specifically for students working above grade level who need a challenging project to complete the unit.

Accommodations may be needed for struggling students, as some of the concepts may be advanced for students who have not taken a business class or have not had any exposure to business concepts. There may be free online versions of this game available for your students.

Launch

The goal of this project is to help students understand how businesses calculate a break-even point and the importance of making a profit. Discuss the concept of breaking even. Ask students about revenue and expenses. It may be helpful to draw a graph with a revenue line and an expense line and talk about the intersection point. It will also be necessary to discuss the difference between fixed costs and variable costs. Students should understand that fixed costs will occur regardless of whether products/services are sold. Variable costs will increase as the number of units/services sold increases.

It may be necessary to brainstorm small business ideas. It will be easiest if students choose a business with one product/service so that product mix is not an issue. Possible small businesses could include the following:

- lemonade stand,
- lawn service,
- coffee shop,
- grocery delivery,
- ice cream stand,
- cupcake stand,
- tutoring,
- sports camp,
- house sitting,
- pet day care,
- babysitting, and
- growing strawberries/raspberries for a "u-pick-it" fruit stand.

Lesson

The student worksheet will lead students through things that need to be considered before they start their small business. After completion of the questions on the worksheet, have students review their answers with a business consultant (adult, teacher, or peer). It may be helpful to limit the time the business will be in operation (e.g., a month, a summer, one year).

Students who choose a business that involves selling a product should research online how much the product costs to purchase and how much the product should be sold for in order to make a profit. The break-even analysis will show how many units of the product will need to be sold.

Students will also have to consider the expenses of doing business. These expenses could include product costs, gas, rent, utilities, wages, advertising, and supplies, among others.

Students who choose a business that is selling a service should research online how much they can charge for their service and then also consider the expenses of doing business. These expenses would be similar to those previously mentioned.

Students should graph a line that represents their sales (sales price x quantity sold) and a line that represents their expenses (fixed expenses and variable expenses). Students could use software like Microsoft Excel to make tables of data and then generate the graph. Have students shade the area where expenses are less than revenue and their business is making money. Then have them shade the area where the expenses are more than revenue and their business is losing money.

Summary

Students should share their business ideas and results with a small group of peers and peer-review their data and graphs. Oral presentations are also an option.

Discussion Questions

➤ What types of businesses have the highest expenses?

➤ What makes a business successful?

➤ What types of businesses are more seasonal?

➤ What types of businesses are affected by the weather?

➤ Where on the graph would you shade to reflect when the business's expenses are higher than the revenue (business is losing money)?

➤ Where on the graph would you shade to reflect when the business's income is higher than the expenses (business is making money)?

➤ What type of training or education is necessary to successfully operate the type of business you chose?

➤ What qualities are necessary for small business owners?

➤ How does the price you charge determine the success of your business?

Systems of Linear Equations

Lesson 6: Wrap Up and Assessment
Small Business Plan

Congratulations! You are opening your own small business. Because you are the business owner, you will get to keep your profits. However, you could also lose money if your business is not successful. To increase your chances of success, you should choose a business that sells a product or service that interests you. For this project, choose the length of time your business will be operating.

In order to get started, there are some things you need to think about and some decisions you need to make. Define the following words, answer the following questions, and then discuss your answers with a business consultant (parent, teacher, or classmate) to help ensure you have thought of everything.

Vocabulary

Entrepreneur
Variable costs
Fixed costs
Customer
Competitors

Questions

1. What is your business name?

2. What product/services will be sold? How many will you sell?

3. Who are your customers?

4. Who are your competitors?

5. What price can you charge your customers for your product/service? How does your price compare with your competitors' prices?

6. What expenses will you incur (fixed and variable)? List the supplies you will need to purchase to start up your business and then keep it running.

7. Where will you advertise?

8. Will weather impact your business?

9. Where will you set up your business? Will you have to pay rent?

Lesson 6: Wrap Up and Assessment

Exit Slip

1. List three things you learned in this unit that you understand and could explain to another student.

2. List three things that are still challenging for you regarding this unit.

Systems of Linear Equations

Unit 3

Exponent Rules and Exponential Functions

In this unit, students will continue their study of functions by identifying exponential growth and decay situations in tables, graphs, and equations. Real-life applications introduced through story problems are included in most lessons. The final project, which could be used as an authentic assessment, is a children's book that students will write to demonstrate their understanding of exponential functions.

This unit begins with a preassessment and four real-life applications of exponential functions that can be discussed in small groups and then as a larger group. Many of the activities will offer the students an opportunity to choose learning activities according to their learning style, personal interests, and readiness level.

What Do We Want Students to Know?

Common Core State Standards Addressed:	Big Ideas
• 8.EE.1 • A.CED.1 • A.SSE.3c • F.IF.7e, 8b • F.LE.1a, 1c, 2, 5	• There are exponent rules that allow us to solve complicated-looking expressions. • Exponential functions are different from linear equations. • Exponential functions always have a variable in the exponent. • The base in the equation determines whether the function will grow or decay. • There are many real-life situations that can be modeled with exponential functions.
	Essential Questions
	• What are the exponent rules and how do I use them? • How are linear and exponential functions the same and different? • How do the base and the exponent affect the shape of the graph? • In real-world situations, how do I know what information goes into the equation and where does it go?

Critical Vocabulary

Base   Exponent   Exponential growth
Exponential decay   Compound growth   Growth factor
Standard form   Initial amount   Decay factor

Unit Objectives

As a result of this unit, students will know:

➤ how to apply the exponent rules to simplify expressions,
➤ that exponential functions have a variable as the exponent,
➤ that exponential functions have growth and decay factors or multipliers,
➤ that exponential growth functions approach zero as x-values decrease, and
➤ that exponential decay functions approach zero as x-values increase.

As a result of this unit, students will understand that:

➤ many real-life situations are modeled in exponential functions, and
➤ for an equation in the form of $y = a(b)^x$, a represents the starting value, the value of b reflects either exponential decay or growth, and x is the time period.

As a result of this unit, students will be able to:

➤ solve exponent functions and simplify expressions by applying the exponent rules;
➤ recognize and describe relationships in which variables grow and decay exponentially;
➤ describe how the values of a and b affect the graph of an equation in the form of $y = a(b)^x$;
➤ recognize exponential relationships in tables, graphs, and equations; and
➤ determine the growth and decay rates in exponential situations.

Launch Scenarios

➤ After Erin graduated from college, she bought a house for $210,000. If it is estimated that real estate is appreciating in value by 5% per year, how much will the house be worth in 10 years when she plans to sell it? (Lesson 2)
➤ When Kevin's daughter is born, he and his wife invest $10,000 in an interest-bearing account for their daughter's education. If the account is earning 4% interest compounded quarterly, and no additional money is

deposited into the account, how long will it take for the account to have $15,000? (Lesson 2)

➤ Katie bought a new car for $30,000. If its value is decreasing by 8% each year, how much will the truck be worth after 3 years? (Lesson 3)

➤ It is estimated that the human body can reduce caffeine in the bloodstream at a rate of 15% per hour. If Zach drinks a cola drink that contains 35 mg of caffeine at 9:00 p.m., how much caffeine will still be his bloodstream at midnight? (Lesson 3)

Unit Overview: Exponent Rules and Exponential Functions

(Assumes a 50–60 minute class period)

Lesson	Whole-Class/Small-Group Lesson Discussion/Activities	Individualized Learning Activities
Preassessment (1/2 period)	Give the preassessment at least several days prior to beginning the unit (20 minutes)	
Lesson 1: Introduction (1–2 periods)	Share results of the preassessment (20 minutes) Introduce launch scenarios and discuss possible methods of solving (10 minutes)	
		Learning Targets and Study Guide (15 minutes) Vocabulary Choice Board (15 minutes)
	Concentration—Review of Exponents, Square Roots, and Integer Multiplication (20 minutes)	
		Exit Slip (10–15 minutes)

Lesson	Whole-Class/Small-Group Lesson Discussion/Activities	Individualized Learning Activities
Lesson 2: Exponential Growth (2–3 periods)	Read *One Grain of Rice* by Demi (1997) or discuss "Doubling Pennies" Complete activity sheet (30 minutes) Hold a group discussion on launch scenarios (15 minutes) Hold a classroom discussion on exponential growth functions (15 minutes)	
		Agenda (30 minutes) Tic-Tac-Toe Board (30 minutes) Exit Slip (10–15 minutes)
Lesson 3: Exponential Decay (2–3 periods)	Hold a class discussion on exponential decay functions (20 minutes)	
		Agenda (30 minutes) Tic-Tac-Toe Board (40 minutes) Think Dots (40 minutes) Exit Slip (5–10 minutes)
Lesson 4: Multiplication Properties of Exponents (2.5 periods)	Hold a class discussion on multiplication properties of exponents (20 minutes)	
		Agenda (30 minutes) Hexagon Puzzles (20 minutes) Exit Slip (5–10 minutes)

Lesson	Whole-Class/Small-Group Lesson Discussion/Activities	Individualized Learning Activities
Lesson 5: Zero and Negative Exponents (2 periods)	Hold a class discussion on zero and negative properties of exponents (20 minutes)	
		Agenda (30 minutes) Tic-Tac-Toe Board (25 minutes) Exit Slip (15 minutes)
Lesson 6: Division Properties of Exponents (2 periods)	Hold a class discussion on division properties of exponents (20 minutes)	
		Agenda (30 minutes) Hexagon Puzzles (20 minutes) Exit Slip (15 minutes)
Lesson 7: Wrap Up and Assessment (3–4 periods)	Collect learning targets and study guide and vocabulary choice board projects	
		Children's Book (40 minutes and outside class time) Exit Slip (10 minutes)

Exponent Rules and Exponential Functions

Preassessment

Directions: Solve the following problems to the best of your ability, and then rate your confidence in your answer in the space provided. Skip questions you do not know. This will not be graded.

1. Identify the following equations as linear, exponential, or neither.

   a. $y = 2x + 5$ _____   c. $y = 2^x + 5$ _____

   b. $y = x^2 + 5$ _____   d. $2y = 5^x - 3$ _____

   _____ I'm sure

   _____ I'm not sure

2. Do the following relationships represent an exponential function?

   a.
x	−1	0	1	2	3	4	5
y	0.5	1	2	4	8	16	32

   b.
x	0	1	2	3	4	5	6
y	1	3	6	12	24	36	48

   _____ I'm sure

   _____ I'm not sure

3. Sketch a graph showing exponential decay and a graph showing exponential growth.

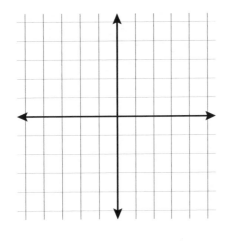 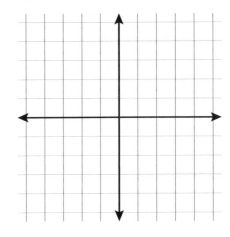

   _____ I'm sure

   _____ I'm not sure

4. Use the multiplication properties of exponents to simplify the following expressions.

a. $2^2 \cdot 2^3$ _____

d. $(4^3)^2$ _____

b. $5^4 \cdot 5^5$ _____

e. $(d^5)^3$ _____

c. $a^3 \cdot a^2$ _____

_____ I'm sure

_____ I'm not sure

5. Use the division properties of exponents to simplify the following expressions.

a. $\dfrac{6^4}{6^2}$ _____

d. $\left(\dfrac{3x}{4y}\right)^2$ _____

b. $\dfrac{a^6}{a^3}$ _____

e. $\left(\dfrac{5}{t}\right)^4$ _____

c. $\left(\dfrac{2}{4}\right)^2$ _____

_____ I'm sure

_____ I'm not sure

6. Use the zero and negative properties of exponents to simplify the following expressions.

a. 2^0 _____

d. $\left(4^{-3}\right)^2$ _____

b. $a^0 \cdot 2^4$ _____

e. $(2d)^{-5}$ _____

c. 3^{-3} _____

_____ I'm sure

_____ I'm not sure

7. For the equation $y = 450(1.08)^4$ identify the starting value, the growth factor, the time period, and the final value.

a. Starting value: _____

c. Time period: _____

b. Growth factor: _____

d. Final value: _____

_____ I'm sure

_____ I'm not sure

8. Write an exponential function to model the following situation and then answer the question. Find the account balance after 3 years if you deposit $650 in an account that pays 4% annual interest compounded annually. _____

_____ I'm sure

_____ I'm not sure

Differentiating Instruction in Algebra 1 © Prufrock Press Inc. • Permission is granted to photocopy or reproduce this page for single classroom use only.

Exponent Rules and Exponential Functions

Learning Targets and Study Guide

At the end of this unit, every student should be able to say:

Learning Target	Explain and Give an Example
I can determine if a table of data represents an exponential relationship. Initial here when mastered _____	
I can determine if an equation represents an exponential relationship. Initial here when mastered _____	
I can sketch graphs showing an exponential growth and an exponential decay situation and describe the situations. Initial here when mastered _____	
I can use the multiplication property and the power to a power property of exponents. Initial here when mastered _____	
I can use the zero and negative exponent properties. Initial here when mastered _____	
I can use the division properties of exponents. Initial here when mastered _____	
I can identify the initial value and the growth/decay factor in an exponential equation. Initial here when mastered _____	
I can solve a real-life story problem involving an exponential function. Initial here when mastered _____	

Exponent Rules and Exponential Functions

Exponent Rules and Exponential Functions

Vocabulary Choice Board

Directions: Using the critical vocabulary for the unit, choose one of the following methods to demonstrate your knowledge of the mathematical terms.

LINCS Cards	Graphic Organizer	News Broadcast
Prepare a LINCS card for each vocabulary word to demonstrate your understanding of the vocabulary words. *(logical–mathematical)*	Construct a graphic organizer of your choice to define each word and demonstrate your understanding of the vocabulary words and how they fit into the current unit. *(verbal/language–mathematical)*	Write a news broadcast that defines all of the vocabulary words, and provide stories that demonstrate your understanding of the vocabulary words. Turn in your news broadcast, and then present it to the class. *(intrapersonal–language)*
Nature Survival	**Creative Story**	**Poster Board**
Prepare a nature survival guide about how each of the vocabulary words would be used if you were stranded on a deserted island. The guide must demonstrate your understanding of the vocabulary words. *(natural–language)*	Write a creative story with color illustrations that demonstrates your understanding of the vocabulary words. *(verbal–language)*	Prepare an 11" x 14" poster board that contains the definitions of the vocabulary words and includes one real-life example for each word that demonstrates your understanding of the definition. *(language–spatial)*
Cartoon	**Rap or Song**	**Acting Out**
Prepare a cartoon strip with at least five frames that defines each word and is illustrated with appropriate pictures to demonstrate your understanding of the vocabulary words. *(visual–spatial)*	Write a rap or a song containing all of the vocabulary words and their correct definitions. Perform your rap or song for the class. *(musical–language)*	Write a script and then act out the words with their correct definitions. Use props if appropriate. *(kinesthetic–language)*

Critical Vocabulary

Base	Exponent	Exponential growth
Exponential decay	Compound growth	Growth factor
Standard form	Initial amount	Decay factor

Exponent Rules and Exponential Functions
Vocabulary Choice Board Rubric

Grading scale:	3 Well done	2 Close to requirements		1 Needs more work
Choice	**Definitions**	**Illustrations and/ or creativity**	**Neatness**	**Other**
LINCS cards	Definitions demonstrate knowledge of vocabulary. Score _____	The illustrations are appropriate and creative. Score _____	The cards are easy to read. Score _____	Semantic connections between words are demonstrated. Score _____
Graphic organizer	Definitions demonstrate knowledge of vocabulary. Score _____	The graphic organizer creatively displays the definitions. Score _____	The graphic organizer is easy to read. Score _____	The graphic organizer choice is appropriate for this assignment. Score _____
News broadcast	Definitions demonstrate knowledge of vocabulary. Score _____	The news broadcast is creative and holds the attention of the audience. Score _____	The news broadcast is easy to read. Score _____	Good presentation skills are demonstrated. Score _____
Nature survival guide	Definitions demonstrate knowledge of vocabulary. Score _____	The guide contains illustrations that are appropriate. Score _____	The guide is easy to read. Score _____	The final product resembles a survival guide. Score _____
Creative story	Definitions demonstrate knowledge of vocabulary. Score _____	The story contains illustrations that are appropriate. Score _____	The story is easy to read. Score _____	The story is interesting. Score _____
Poster board	Definitions demonstrate knowledge of vocabulary. Score _____	The poster board includes illustrations that are appropriate. Score _____	The poster board is easy to read. Score _____	Real-life examples are included for each vocabulary term. Score _____
Cartoon	Definitions demonstrate knowledge of vocabulary. Score _____	The cartoon includes illustrations that are appropriate. Score _____	The cartoon is easy to read. Score _____	The cartoon contains a minimum of five frames. Score _____
Rap or song	Definitions demonstrate knowledge of vocabulary. Score _____	The rap or song is creative. Score _____	The lyrics are easy to read. Score _____	Good performance skills are demonstrated. Score _____
Acting out	Definitions demonstrate knowledge of vocabulary. Score _____	The script is creative. Score _____	The script is easy to read. Score _____	Good performance skills are demonstrated. Score _____
				Total: _____/12= _____%

Comments:

Lesson 1: Introduction

Concentration: Review of Exponents, Square Roots, and Integer Multiplication

Students can work in pairs to play this review game. There are two concentration boards with the same information, just in a different order (see the next two pages). Have one pair of students cover all of the spaces on one board with precut pieces of paper, and then give the board to a group that has the opposite board. This eliminates students from seeing the answers before they start the game. Students can put their initials on the squares they win to determine who finds the most pairs.

Concentration Board

$\sqrt{25}$	27	2^2	$-9\cdot-3$	10^2	$-2\cdot6$	$\sqrt{36}$	49
25	3	7	80	$b\cdot b$	5	d^2	16
$d\cdot d$	4^2	$\sqrt{16}$	-12	6	100	5^2	4
$4\cdot-4$	$\sqrt{49}$	4	8	9^2	$\sqrt{64}$	$\sqrt{4}$	9
$\sqrt{1}$	1	$2\cdot-5$	10	$\sqrt{81}$	-16	b^2	36
-18	7^2	$9\cdot-2$	81	2	$6\cdot-4$	8^2	-24
$-8\cdot-10$	$\sqrt{100}$	64	3^2	9	-10	6^2	$\sqrt{9}$

Concentration Board

$\sqrt{64}$	9	5^2	$4 \cdot -4$	6^2	$9 \cdot -2$	$\sqrt{16}$	6
8	2	10	1	$d \cdot d$	-16	b^2	81
$b \cdot b$	9^2	$\sqrt{9}$	-24	36	9	3^2	3
$6 \cdot -4$	$\sqrt{81}$	80	64	d^2	$\sqrt{36}$	$\sqrt{25}$	7
$\sqrt{100}$	4	$-8 \cdot -10$	4	$\sqrt{49}$	-18	2^2	5
-12	10^2	$2 \cdot -5$	16	49	$6 \cdot -2$	7^2	-10
$-9 \cdot -3$	$\sqrt{1}$	100	8^2	27	25	4^2	$\sqrt{4}$

Lesson 1: Introduction

Exit Slip

1. Explain how square roots and squaring numbers are related. Demonstrate your answer using an example.

2. Determine the pattern in each of the following lists of numbers, and then complete the sequences.

   a. $\{2, 4, 6, 8, \underline{\hspace{1.2cm}}, \underline{\hspace{1.2cm}}, \underline{\hspace{1.2cm}}\}$

   b. $\{2, 8, 32, 128, \underline{\hspace{1.2cm}}, \underline{\hspace{1.2cm}}, \underline{\hspace{1.2cm}}, \underline{\hspace{1.2cm}}\}$

   c. $\{1000000, 250000, 62500, 15625, \underline{\hspace{2cm}}, \underline{\hspace{2cm}}\}$

   d. $\{500, 50, 5, 0.5, \underline{\hspace{1.2cm}}, \underline{\hspace{1.2cm}}, \underline{\hspace{0.6cm}} . \underline{\hspace{0.6cm}}\}$

Lesson 2: Exponential Growth

One Grain of Rice by Demi and Doubling Pennies Lesson Plan

There are several children's books that demonstrate the concept of exponential growth. *One Grain of Rice* by Demi is easy to read to students of any age and provides an opportunity for learners of all styles to understand this concept. Another well-known application of exponential growth is the story of a penny a day. However, the title "A Penny a Day" is misleading. There are blogs that dispute what the name of the actual lesson plan should be. I will call it "Doubling Pennies." Both of these examples simulate exponential growth in terms that most students can understand and may find interesting.

Launch

One Grain of Rice	**Doubling Pennies**
• Ask students if they are familiar with the story of the village girl named Rami who is able to trick the Raja into giving rice to the villagers who are starving due to a famine. • This is an opportunity to talk about famine and other social issues that students are learning about in their social studies or history classes.	• Ask students if they were given the choice of receiving one million dollars 30 days from now or a penny given on the first day and the amount doubled every day after that for 30 days, which option they would choose. • This may be the time to clarify how the doubling will take place. One penny the first day, 2 pennies the second day, 4 pennies the third day, 8 pennies the fourth day, and so on. Keep in mind that the money accumulates each day, so after the first day there is one penny, after the second day the balance is 3 pennies, after the third day there are 7 pennies, and so on.

Lesson

One Grain of Rice	**Doubling Pennies**
• The story is nicely divided by days, which make it easy to stop reading and have the students record the number of grains of rice after each day. • Check for understanding and make sure the students' calculations are correct. • Have students write in words how each day's rice amount is calculated.	• After students have chosen an option, assign them to small groups and have them briefly discuss the option they chose and why. • Students can complete the tracking sheet either in small groups, independently, or as a class. • Have students write in words how the number of pennies is calculated each day.

Differentiating Instruction in Algebra 1 © Prufrock Press Inc. • Permission is granted to photocopy or reproduce this page for single classroom use only.

Summary

One Grain of Rice	Doubling Pennies
• Students should complete the table of data. • For the first 5 days, have students calculate the number of grains of rice that have accumulated after each day and write that number in the corner of each day's square. Have students write how to calculate this cumulative amount of rice. • Using the two written descriptions, have students write an equation that models the growth in rice. Remind them that this equation will be used to find the amount of rice received on the *n*th day.	• Students should complete the table of data. • For the first 5 days, have students calculate the number of pennies that have accumulated after each day and write that number in the corner of each day's square. Have students write how to calculate this cumulative amount of pennies. • Using the two written descriptions, have students write an equation that models the growth in pennies. Remind them that this equation will be used to find the number of pennies received on the *n*th day.

Discussion Questions

➤ What is the equation that models this situation?
➤ How much rice or how many pennies is/are received on the 40th day? The 50th day?
➤ What other everyday situations demonstrate exponential growth?

Lesson 2: Exponential Growth

One Grain of Rice by Demi and
Doubling Pennies Tracking Sheet

Directions

➤ For the first 5 days, write in small numbers in a corner the accumulated amount on each day.

➤ Write in words how the amounts for each day are calculated.

➤ Write in words how the cumulative totals for each day are calculated.

➤ Write an equation that models this situation using the descriptions from the second and third steps above.

Day 1	Day 2	Day 3	Day 4	Day 5	Total after 5 days
Day 6	**Day 7**	**Day 8**	**Day 9**	**Day 10**	**Total after 10 days**
Day 11	**Day 12**	**Day 13**	**Day 14**	**Day 15**	**Total after 15 days**
Day 16	**Day 17**	**Day 18**	**Day 19**	**Day 20**	**Total after 20 days**
Day 21	**Day 22**	**Day 23**	**Day 24**	**Day 25**	**Total after 25 days**
Day 26	**Day 27**	**Day 28**	**Day 29**	**Day 30**	**Total after 30 days**

Exponent Rules and Exponential Functions

Lesson 2: Exponential Growth

Agenda

Imperatives (You must do all three of these.)

1. Determine the pattern in each of the following lists of numbers. Complete the following sequences, and then finish writing the rule that describes the sequence.

   a. 1, 2, 4, 8, _____, _____, _____

   Start with 1 and _____

   b. 1, 3, 9, 27, _____, _____, _____

   Start with 1 and _____

   c. 1, 10, 100, _____, _____, _____

   Start with 1 and _____

   d. 4, 10, 25, _____, _____, _____

   Start with _____ and _____

   e. 6, 7.5, 9.375, _____, _____, _____

   Start with _____ and _____

2. Write the following percentages as decimals.

   a. 25%   _____   g. 10%   _____

   b. 12%   _____   h. 8%   _____

   c. 36%   _____   i. 1%   _____

   d. 100%   _____   j. 0.10%   _____

   e. 102%   _____   k. 0.25%   _____

   f. 200%   _____   l. 0.500%   _____

3. Tell whether the each set of ordered pairs or table values satisfies an exponential function. If it does, what is the growth rate?

   a. $(-3,-8),(-2,-5),(-1,-2),(0,1),(1,4),(2,7)$

   b. $(-2,\frac{1}{9}),(-1,\frac{1}{3}),(0,1),(1,3),(2,9),(3,27)$

   c.

x	−3	−2	−1	0	1	2	3	4	5
y	.125	.25	.5	1	2	4	8	16	32

Negotiables (You must do at least two of these.)

1. Evaluate each exponential function, and then answer the questions.
   a. The function $f(t) = 3(2)^t$ models a bacteria population after t hours.

      What will be the population after 12 hours? _____

      According to the equation, what is the initial number of bacteria? _____

      What is the growth rate? _____

   b. A rabbit population can be modeled with the function $f(t) = 18(3)^t$ where t represents years.

      What will be the population after 5 years? _____

      According to the equation, what is the initial number of rabbits? _____

      What is the growth rate? _____

2. Write an exponential function to represent the following situation, and then answer the question.

   After Erin graduated from college, she bought a house for $210,000. If it is estimated that real estate is appreciating in value by 5% per year, how much will the house be worth in 10 years when she plans to sell it?

3. Write an exponential function to represent the following situation, and then answer the question.

   When Kevin's daughter is born, he and his wife invest $10,000 in an interest-bearing account for their daughter's education. If the account is earning 4% interest compounded annually, and no additional money is deposited into the account, how long will it take for the account to have $15,000?

Options (You may choose to do these problems.)

1. Which grows faster as x is increasing, 4^x or x^4? Explain how you made your decision.

2. Using $x = 10$, evaluate both expressions and determine the difference in the two values.

Exponent Rules and Exponential Functions

Lesson 2: Exponential Growth
Tic-Tac-Toe Board

Directions: Choose three problems in a row, in a column, or on a diagonal. Solve the problems and show your work on a separate sheet of paper. Circle the problems you choose on this page and then reference them on your work paper with the number from the square.

1	**2**	**3**
Identify the equations as linear, exponential, or neither. Identify the slope for the linear equations and the rate of growth for exponential equations. a. $3y = 4x - 12$ b. $y = 2^x$ c. $x^2 = 5y - 9$ d. $y = 10(1 + .2)^x$ e. $2y = 5^x$	Carol deposits \$1,000 in an account that pays interest compounded yearly. If the money is left in the account for 10 years, how much will be in the account for the following interest rates? a. 4% b. 7% c. 10%	Determine which of the following functions represent growth models. For those functions, determine the growth rate. a. $y = 3(1.9)^x$ b. $y = .75(1.1)^x$ c. $y = 5(0.75)^x$ d. $y = (0.98)^x$ e. $y = 3(1)^x$
4	**5**	**6**
Describe how to calculate the growth factor of an exponential function from a table. 	x	y
---	---
−1	2.5
0	10
1	40
2	160
3	640		If a prairie dog population is tripling in size every year, what is the growth factor? Also express it as a percentage. How many prairie dogs would there be after 6 years if the initial population was 16?	Tom deposits \$1,500 in an account that pays 5% interest compounded yearly. Find the balance in the account for the given time periods. a. 5 years b. 9 years c. 15 years
7	**8**	**9**
In 2010, an antique car is worth \$25,000. If it is increasing in value by 6% each year, how much will the car be worth if you keep it for 6 years? How long will it take for the car to have a value of \$50,000?	Which exponential function will grow faster? Explain your answer. $$y = 10(1.10)^x$$ $$y = 3(1 + .2)^x$$ Which of these functions has the highest value when $x = 10$?	Fill in the missing values in the table to represent an exponential function. 	x	y
---	---
−1	2.5
0	5
1	
2	20
3	40
4		

Lesson 2: Exponential Growth
Exit Slip

1. Write the general equation representing exponential growth. Define all variables in the equation.

2. Jackie deposits $5,600 in an account earning 2.5% interest compounded yearly. How long will it take for the account to have at least $6,500? Write an equation and then use it to answer the question.

Lesson 3: Exponential Decay
Agenda

Imperatives (You must do all three of these.)

1. Determine the pattern in each of the following lists of numbers. Complete the following sequences and finish writing the rule that describes the sequence. In the last problem, generate your own sequence and then write the rule.

   a. 24, 12, 6, _____, _____, _____

   Start with 24 and _____

   b. 100, 10, 1, _____, _____, _____

   Start with 100 and _____

   c. 200, 50, 12.5, _____, _____, _____

   Start with 200 and _____

   d. 72, 18, 4.5, _____, _____, _____

   Start with _____

   e. _____, _____, _____, _____, _____, _____

   Start with _____

2. Write the following percentages as decimals in the form of "1 minus a number." The first one is completed as an example.

25%	12%	36%	100%	91%	86%
$(1-.75)$					
10%	8%	1%	3%	65%	99% .

3. Tell whether the each set of ordered pairs or table of values satisfies an exponential function. If it does, what is the decay factor?

   a. $(-3, 960), (-2, 480), (-1, 240), (0, 120), (1, 60), (2, 30)$

   b. $(-2, 220), (-1, 170), (0, 120), (1, 70), (2, 20), (3, -30)$

   c.
x	−1	9	1	2	3	4
y	5,000	500	50	5	0.5	0.05

Negotiables (You must do at least two of these.)

1. Evaluate each exponential function, and then answer the questions.

    a. The function $f(t) = 1,000(1 - .20)^t$ models the value of an HD television after t years.

       What will be the value after 4 years? _____

       According to the equation, what is the initial cost of the television? _____

       What is the decay rate? _____

    b. A fish population decreasing in a lake due to environmental pollution can be modeled with the function $f(t) = 1,500(.65)^t$ where t represents years.

       What will be the population after 3 years? _____

       According to the equation, what is the initial number of fish in the lake? _____

       What is the decay rate? _____

2. Write an exponential function to represent the following situation, and then answer the question.

    Katie bought a new car for $30,000. If its value is decreasing by 12% each year, how much will the car be worth after 3 years?

3. Write an exponential function to represent the following situation, and then answer the question.

    It is estimated that the human body can reduce caffeine in the bloodstream at a rate of 15% per hour. If Zach drinks a cola drink that contains 35 mg of caffeine at 9:00 p.m., how much caffeine will still be his bloodstream at midnight?

Options (You may choose to do this problem.)

1. Graph $y = 10(.75)^x$ and $y = -10(.75)^x$ on your calculator and then sketch them below. For both equations, describe what is happening to the y-values as the x-values increase.

Lesson 3: Exponential Decay
Tic-Tac-Toe Board

Directions: Choose three problems in a row, in a column, or on a diagonal. Solve the problems and show your work on a separate sheet of paper. Circle the problems you choose on this page and then reference them on your work paper with the number from the square.

1	**2**	**3**
Describe real-life situations that can be modeled with exponential functions. There should be two for growth models and two for decay models.	A truck is purchased for $25,000 and decreases in value 12% each year. Write an exponential equation to model this situation and then determine when the value of the truck will be less than $10,000.	Determine which of the following functions represent decay models. For those functions, determine the decay rate. a. $y = 2(1.2)^x$ b. $y = .75(0.9)^x$ c. $y = 5(1.75)^x$ d. $y = (0.97)^x$ e. $y = 3(.01)^x$
4	**5**	**6**
Describe how to calculate the decay factor of an exponential function from a table. $\begin{array}{c\|c} x & y \\ \hline -1 & 1{,}111 \\ 0 & 1{,}000 \\ 1 & 900 \\ 2 & 810 \\ 3 & 729 \end{array}$	Which exponential function will decay faster? Explain your answer. $$y = 10(0.80)^x$$ $$y = 9(1 - .25)^x$$ Considering the equation $y = a(b)^x$, what values of b will result in exponential decay?	A bouncy ball is dropped from a height of 95 cm and rebounds 75% of the height on each bounce. Write a function to represent the function and determine after how many bounces the ball would rebound to a height less than 10cm.
7	**8**	**9**
Monique is a contestant on a game show. She starts with one million dollars. Every time she answers a question incorrectly, half of the money is taken away. If she answers the first three questions wrong, how much money will she still have?	In 2010, Small Town's population was 1,759 people. If it is declining by 2% each year, what will be the population in 2020? How long until the population is less than 1,000?	Fill in the missing value in the table to represent an exponential function. $\begin{array}{c\|c} x & y \\ \hline -1 & 500 \\ 0 & 100 \\ 1 & 20 \\ 2 & \\ 3 & 0.8 \\ 4 & 0.16 \end{array}$

Lesson 3: Exponential Decay
Think Dots

Directions: In your groups, each student will take turns rolling the die. Answer the corresponding question on the record sheet. You may discuss your answers with your group members before recording your answer. You must answer all of the questions.

⚀	⚁	⚂
Write a decrease of 25% and an increase of 50% two different ways. Think about what decay and growth equations look like.	Draw a picture of a decay function. Write three equations that represent decay functions.	For an exponential function in the form of $f(x) = A(1+r)^x$, explain what the terms $f(x)$, A, $1+r$, and x represent.
⚃	⚄	⚅
Write exponential functions for the following scenarios: a. Starts at 250 and increases at a rate of 6% b. Starts at 5 and decreases by 2% c. Starts at 300 and increases by 100%	Identify each of the exponential equations as growth or decay. a. $y = 2.5^x$ b. $y = 0.5^x$ c. $y = (1-.25)^x$ d. $y = (1+.45)^x$	A standard-size piece of paper is folded in half and then in half again, multiple times. Complete the table showing the area of the paper after each fold (round to two decimals). <table><tr><td>x</td><td>y</td></tr><tr><td>0</td><td>93.5</td></tr><tr><td>1</td><td></td></tr><tr><td>2</td><td></td></tr><tr><td>3</td><td></td></tr><tr><td>4</td><td></td></tr></table>

Exponent Rules and Exponential Functions

Lesson 3: Exponential Decay
Think Dots

Directions: In your groups, each student will take turns rolling the die. Answer the corresponding question on the record sheet. You may discuss your answers with your group members before recording your answer. You must answer all of the questions.

Exponent Rules and Exponential Functions

Find the coordinates of a point that can be found on both of the following functions. $$f(x) = (.75)^x$$ $$f(x) = (1.25)^x$$	Explain why exponential decay functions approach zero as the x-values increase but they never reach zero. Draw a picture of a decay function to help explain your position.	A car is worth $25,000 today, and it is declining in value by 10% per year. Explain what $f(0)$ and $f(-1)$ mean in the context of this situation.
		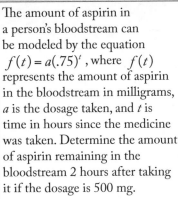
The amount of aspirin in a person's bloodstream can be modeled by the equation $f(t) = a(.75)^t$, where $f(t)$ represents the amount of aspirin in the bloodstream in milligrams, a is the dosage taken, and t is time in hours since the medicine was taken. Determine the amount of aspirin remaining in the bloodstream 2 hours after taking it if the dosage is 500 mg.	Explain why the function $f(x) = 80(1 + .25)^x$ could not be used to represent a bouncy ball being dropped from a height of 80 cm. Describe a scenario that could be represented with this equation.	Describe a real-life situation that could be modeled with each of the following functions. a. $f(t) = 600(1 + 0.5)^t$ b. $f(t) = 40(1 - .5)^t$

Differentiating Instruction in Algebra 1 © Prufrock Press Inc. • Permission is granted to photocopy or reproduce this page for single classroom use only.

Lesson 3: Exponential Decay
Think Dots

Directions: In your groups, each student will take turns rolling the die. Answer the corresponding question on the record sheet. You may discuss your answers with your group members before recording your answer. You must answer all of the questions.

Explain why exponential decay functions approach zero as the *x*-values increase but they never reach zero. Draw a picture of a decay function to help explain your position.	A car is worth $30,000 today and it is declining in value by 8% per year. Explain what $f(0)$ and $f(-1)$ mean in the context of this situation.	Describe a real-life situation that could be modeled with the function $f(t) = 600(1-.25)^t$.
In your first professional job, you are offered two different salary scenarios. You will receive $45,000 the first year and then the second and subsequent years you can either receive $2,000 raises each year or a 4% increase each year from the previous year's salary. Which plan would you choose? Write equations to model both situations. Would it depend on how long you were planning to stay in this position?	For the function $f(x) = 5(2)^x$, what is the range for a domain of $0 < x \le 10$? Your answer should be written in the form of ___ $< y \le$ ___.	Money put into a bank account with interest compounded annually could be modeled with the equation $f(t) = 600(1.06)^t$. How would the equation have to change if the interest were compounded monthly instead of annually? Describe how you would find out how much money would be in the account after 1.5 years using a table or your calculator.

Exponent Rules and Exponential Functions

Exponent Rules and Exponential Functions

Lesson 3: Exponential Decay

Think Dots Record Sheet

Circle worksheet chosen:

Lesson 3: Exponential Decay

Exit Slip

1. Write the general equation representing exponential decay. Define all variables in the equation.

2. Give an everyday example of exponential decay.

Differentiating Instruction in Algebra 1 © Prufrock Press Inc. • Permission is granted to photocopy or reproduce this page for single classroom use only.

Exponent Rules and Exponential Functions

Lesson 4: Multiplication Properties of Exponents

Agenda

Imperatives (You must do all of these.)

1. Simplify the expressions by using exponents.

   a. $a \cdot a$ _____

   b. $b \cdot b \cdot b$ _____

   c. $c \cdot c \cdot c \cdot c$ _____

   d. $2 \cdot a \cdot a$ _____

   e. $4 \cdot c \cdot c \cdot d$ _____

   f. $3 \cdot 3 \cdot 3 \cdot 3$ _____

2. Write each expression as repeated multiplication. The first one is done as an example.

   a. $2^2 \cdot 2^3$ ___$2 \cdot 2 \cdot 2 \cdot 2 \cdot 2$___

   b. $b^3 \cdot b^4$ _____

   c. $a^2 \cdot a^3 \cdot a^2$ _____

   d. $4^3 \cdot 4^2 \cdot 2^3$ _____

3. Simplify the expressions using the multiplication property.

   a. $x^3 \cdot x^4$ _____

   b. $6^2 \cdot 6^3$ _____

   c. $5^2 \cdot 5^3 \cdot 5^6$ _____

   d. $a^3 \cdot b^2 \cdot a^5$ _____

   e. $b^{-3} \cdot b^4$ _____

   f. $4^3 \cdot 2^4 \cdot 2^3 \cdot 4^{-2}$ _____

   g. $a^6 \cdot x^3 \cdot a^{-2} \cdot a$ _____

   h. $y \cdot 4^2 \cdot 4^3$ _____

4. Simplify the expressions using the power properties.

   a. $(x^3)^4$ _____

   b. $(4^2)^3$ _____

   c. $(2^{-2})^{-5}$ _____

   d. $(3x)^3$ _____

   e. $(4x^2)^3$ _____

   f. $(3^2 b^3)^3$ _____

   g. $(3^2 x^2)^3$ _____

   h. $(4^2 2 b^2)^2$ _____

Exponent Rules and Exponential Functions

Negotiables (You must do at least two of these.)

1. Explain the difference between the following two expressions and then solve them:

$$-(2x)^2 \text{ and } (-2x)^2$$

2. Answer the following questions using complete sentences.

   a. Can $x^2 \cdot t^2$ be simplified?

   b. Are $(x^2)^3$ and $x^2 \cdot x^3$ equivalent expressions?

3. A storage box is $3x^2$ wide, x^2 long, and $4x$ high. Write an expression to represent the volume of the box. Draw a three-dimensional sketch of the storage box.

Options (You may choose to do this problem.)

1. Explain the error made in simplifying the following expression. Simplify the expression correctly using multiplication properties of exponents.

$$(a+1)^3 = a^3 + 1^3$$

Lesson 4: Multiplication Properties of Exponents

Hexagon Puzzle

Directions: Solve any problem in a hexagon in the far left-hand column. If you get the answer correct, choose a problem from the next column that is adjacent to the first problem. If you get a problem wrong, move to another problem in the same column adjacent to the hexagon and solve it. Make a path to the right by solving the problems. You must complete a continuous path of problems from left to right.

A → Simplify the expression using exponents.
$$a \cdot a \cdot b \cdot b$$

B → Write each expression as repeated multiplication.
$$2^3 \cdot 2^5$$

C → Simplify the expression using exponents.
$$b \cdot c \cdot b \cdot c$$

D → Write each expression as repeated multiplication.
$$t^2 \cdot r^4$$

E → Simplify the expression.
$$x^2 \cdot x^3$$

F → Simplify the expression.
$$x^2 \cdot x^2$$

G → Simplify the expression.
$$x^4 \cdot 2x^3$$

H → Simplify the expression.
$$t^4 \cdot t^6$$

I → Simplify the expression.
$$x^2 \cdot x^3 \cdot x^2$$

J → Simplify the expression.
$$x^3 \cdot x^2 \cdot t^3 \cdot t^4$$

K → Simplify the expression.
$$2a^3 \cdot b^5 \cdot a^2 \cdot 4b^2$$

L → Simplify the expression.
$$2^3 \cdot 2^5 \cdot 3x^2 \cdot 2^3$$

M → Simplify the expression.
$$(x^2)^2$$

N → Simplify the expression.
$$(b^3)^2$$

O → Simplify the expression.
$$(x^3)^2$$

P → Simplify the expression.
$$(e^4)^2$$

Q → Simplify the expression.
$$(4^3)^3$$

R → Simplify the expression.
$$(3x^3)^4$$

S → Simplify the expression.
$$(x^3 \cdot x^2)^2$$

T → Simplify the expression.
$$(d^3 \cdot d^4)^3$$

U → Simplify the expression.
$$(2x^4 \cdot x^3)^3$$

V → Simplify the expression.
$$(4b^2 \cdot b^2)^3$$

W → Simplify the expression.
$$(c^2 \cdot x^4)^2$$

X → Simplify the expression.
$$(x^4 \cdot x^3)^3$$

Exponent Rules and Exponential Functions

Lesson 4: Multiplication Properties of Exponents

Hexagon Puzzle

Directions: Solve any problem in a hexagon in the far left-hand column. If you get the answer correct, choose a problem from the next column that is adjacent to the first problem. If you get a problem wrong, move to another problem in the same column adjacent to the hexagon and solve it. Make a path to the right by solving the problems. You must complete a continuous path of problems from left to right.

A→Simplify the expression using exponents. $a \cdot a \cdot b \cdot b$

B→Write each expression as repeated multiplication. $2^3 \cdot 2^5$

C→Simplify the expression using exponents. $b \cdot c \cdot b \cdot c$

D→Write each expression as repeated multiplication. $t^3 \cdot r^5$

E→Simplify the expression. $x^3 \cdot x^5$

F→Simplify the expression. $x^2 \cdot x^2$

G→Simplify the expression. $x^4 \cdot x^8$

H→Simplify the expression. $t^4 \cdot t^6$

I→Simplify the expression. $x^2 \cdot x^{-3} \cdot x^2$

J→Simplify the expression. $x^{-3} \cdot x^4 \cdot t^2 \cdot t^6$

K→Simplify the expression. $2a^3 \cdot b^5 \cdot a^2 \cdot 4b^{-2}$

L→Simplify the expression. $-2^3 \cdot 2^5 \cdot 3x^{-2} \cdot 2^6$

M→Simplify the expression. $(x^3)^2$

N→Simplify the expression. $(d^4)^3$

O→Simplify the expression. $(4^3)^3$

P→Simplify the expression. $(3x^3)^4$

Q→Simplify the expression. $(x^3 \cdot x^2)^2$

R→Simplify the expression. $(c^2 \cdot c^4)^3$

S→Simplify the expression. $(2x^4 \cdot x^3)^3$

T→Simplify the expression. $(4b^2 \cdot 2b^{-1})^3$

U→Simplify the expression. $(x^4 \cdot x^3)^3 \cdot x^2$

V→Simplify the expression. $(c^2 \cdot x^4)^2 \cdot x^3$

W→Simplify the expression. $(2c^3 \cdot c^4)^2 \cdot 7c^3$

X→Simplify the expression. $(2c^3) \cdot (3c^4)^2 \cdot (c^{-2})^3$

Exponent Rules and Exponential Functions

Lesson 4: Multiplication Properties of Exponents

Hexagon Puzzle

Directions: Solve any problem in a hexagon in the far left-hand column. If you get the answer correct, choose a problem from the next column that is adjacent to the first problem. If you get a problem wrong, move to another problem in the same column adjacent to the hexagon and solve it. Make a path to the right by solving the problems. You must complete a continuous path of problems from left to right.

A→Simplify the expression.
$2x^{-3} \cdot 5x^5$

B→Simplify the expression.
$3x^7 \cdot -x^{-2}$

C→Simplify the expression.
$-x^4 \cdot -2x^8$

D→Simplify the expression.
$2t^4 \cdot -4x^6$

E→Simplify the expression.
$x^2 \cdot x^{-3} \cdot x^2$

F→Simplify the expression.
$x^{-3} \cdot x^4 \cdot t^2 \cdot t^6$

G→Simplify the expression.
$2a^3 \cdot b^5 \cdot a^2 \cdot 4b^{-2}$

H→Simplify the expression.
$-2^3 \cdot 2^5 \cdot 3x^{-2} \cdot 2^6$

I→Simplify the expression.
$(x^{-4})^{-2}$

J→Simplify the expression.
$(2d^4)^5$

K→Simplify the expression.
$(-4^3)^2$

L→Simplify the expression.
$(-3x^3)^4$

M→Simplify the expression.
$(2x^3 \cdot x^2)^2$

N→Simplify the expression.
$(2c^{-2} \cdot 3c^{-4})^{-3}$

O→Simplify the expression.
$(-2x^3 \cdot -x^2)^5$

P→Simplify the expression.
$(-4b^5 \cdot 2b^{-1})^2$

Q→Simplify the expression.
$(x^3 \cdot x^4)^2 \cdot 2x^4$

R→Simplify the expression.
$(3c^2 \cdot 3x^4)^2 \cdot x^3$

S→Simplify the expression.
$(-2c^3 \cdot c^4)^2 \cdot -7c^{-3}$

T→Simplify the expression.
$(2c^3) \cdot (3c^4)^2 \cdot (c^{-2})^3$

U→Simplify the expression.
$-(2x \cdot x^3)^2$

V→Simplify the expression.
$-(4x \cdot x^4)^2$

W→Simplify the expression.
$-(2x \cdot x^3)^2 \cdot -3x^2$

X→Simplify the expression.
$-(-3x^3 \cdot 2x^2)^2$

Exponent Rules and Exponential Functions

Lesson 4: Multiplication Properties of Exponents

Exit Slip

1. Explain in words the following multiplication properties of exponents:

   a. $a^m \cdot a^n$

   b. $(a^m)^n$

2. Create two new problems for each of the properties above, and then solve the problems.

Lesson 5: Zero and Negative Exponents
Agenda

Imperatives (You must do all of these.)

1. Complete the tables of values. It may be easiest to start at the right and work to the left. Look for patterns in the tables. Use fractions where necessary.

a.

2^{-4}	2^{-3}	2^{-2}	2^{-1}	2^{0}	2^{1}	2^{2}	2^{3}	2^{4}	2^{5}

b.

4^{-4}	4^{-3}	4^{-2}	4^{-1}	4^{0}	4^{1}	4^{2}	4^{3}	4^{4}	4^{5}

2. Evaluate or simplify each expression.

a. a^{0} _____

b. 3^{0} _____

c. $5a^{0}$ _____

d. $(5a)^{0}$ _____

e. $(4^{-2})^{0}$ _____

f. $5^{-1} \cdot a^{0}$ _____

3. Evaluate the exponential expressions. Write fractions in simplest form.

a. 4^{-2} _____

b. 2^{-4} _____

c. $4(2^{-3})$ _____

d. $2^{3} \cdot 2^{-5}$ _____

e. $\left(\dfrac{1}{3}\right)^{-3}$ _____

f. $4^{-3} \cdot 2^{-4} \cdot 2^{3} \cdot 4^{-2}$ _____

g. $5^{-6} \cdot 5^{6}$ _____

h. $10 \cdot 10^{-1}$ _____

4. Simplify by rewriting the expressions with positive exponents.

a. x^{-2} _____

b. y^{-4} _____

c. $4x^{-3}$ _____

d. $2a^{-3}$ _____

e. $\dfrac{2}{t^{-3}}$ _____

f. $\dfrac{m^{-3}}{6}$ _____

g. $x^{4} \cdot y^{-6}$ _____

h. $(-8a)^{0}$ _____

Exponent Rules and Exponential Functions

Negotiables (You must do at least two of these.)

1. Is the expression $2a^{-3}$ equivalent to $\dfrac{1}{2a^3}$? Explain your answer.

2. Write equivalent expressions using positive exponents. Evaluate if possible.

   a. $2a^{-4}$ _____

   b. -3^{-4} _____

   c. $-5x^{-3}$ _____

   d. $(-5a)^{-2}$ _____

3. Explain how the graphs of 2^x and $\left(\dfrac{1}{2}\right)^x$ are similar and different.

Options (You may choose to do this problem.)

1. A bank account currently has a balance of $18,233. If the money was deposited 4 years ago into an account earning 5% interest, how much money was originally deposited?

   a. Write an equation and solve it.

   b. Explain how negative exponents can be used to solve this type of problem.

Lesson 5: Zero and Negative Exponents
Tic-Tac-Toe Board

Directions: Choose three problems in a row, in a column, or on a diagonal. Solve the problems and show your work on a separate sheet of paper. Circle the problems you choose on this page and then reference them on your work paper with the number from the square.

1	**2**	**3**
Evaluate each expression for $x = 1, x = 0,$ and $x = -1$ a. 5^x b. -5^x c. $(-5)^x$	Find the value of x that makes the statements true. a. $\dfrac{1}{3} = 3^x$ b. $\dfrac{1}{16} = x^{-2}$ c. $6^{-2} = \dfrac{1}{x}$	Write a real-life situation that can be modeled by $f(t) = 6,000(1+.03)^t$. Solve it for $t = 0$. Explain what this answer means given the context of your situation.
4	**5**	**6**
An antique car was purchased in 2000 for $20,500. If it is estimated that the car is increasing in value by 3% per year, how much is it worth now? How much was it worth in 2007?	A bank account earns 3% interest compounded yearly. If you want $10,000 in the account in 7 years, approximately how much should you deposit into the account now?	Evaluate each expression for $x = 1, x = 0,$ and $x = -1$ (you may have a b in your answer). a. $5b^x$ b. $-5x^3$ c. $(-5b)^x$
7	**8**	**9**
Evaluate each expression for $x = 2, y = 3, z = -1$. a. x^{-2} b. $(x \cdot z)^{-3}$ c. y^{-x} d. $(xz)^0$	If a car is currently worth $9,000 and it has been decreasing in value by 12% per year, how much was it worth 3 years ago? How much was it worth 7 years ago when it was originally purchased?	Simplify the expressions. a. $\dfrac{a^3}{c^{-2}}$ b. $\dfrac{d^{-3}}{c^2}$ c. $\dfrac{b^{-2}}{a^{-2}}$

Lesson 5: Zero and Negative Exponents
Exit Slip

1. Simplify the following expressions, and then explain in words the zero and negative properties of exponents.

   a. a^0

   b. a^{-1}

2. Create two new problems for each of the above properties, and then solve the problems.

Lesson 6: Division Properties of Exponents

Agenda

Imperatives (You must do all of these.)

1. Write each expression as repeated multiplication, and then eliminate the common factors in the numerator and denominator. The first problem is completed as an example.

    a. $\dfrac{a^5}{a^2}$ $\dfrac{a \cdot a \cdot a \cdot a \cdot a}{a \cdot a}$ _____

    c. $\dfrac{6^2}{6^4}$ _____

    b. $\dfrac{3^4}{3^3}$ _____

    d. $\dfrac{a^4 c^3}{a^3 c^2}$ _____

2. Simplify the expressions, and reduce fractions to lowest terms.

    a. $\left(\dfrac{3}{4}\right)^3$ _____

    c. $\left(\dfrac{3}{2}\right)^4$ _____

    b. $\left(\dfrac{3}{y}\right)^4$ _____

    d. $\left(\dfrac{2}{4}\right)^2$ _____

3. Simplify the expressions, and reduce fractions to lowest terms.

    a. $\left(\dfrac{x^3}{x^2}\right)^2$ _____

    e. $\left(\dfrac{2x}{3y}\right)^0$ _____

    b. $\left(\dfrac{2x}{4y}\right)^2$ _____

    f. $\left(\dfrac{2x^2}{4y^3}\right)^2$ _____

    c. $\left(\dfrac{5w}{3w}\right)^3$ _____

    g. $\left(\dfrac{2x^3}{x^2}\right)^5$ _____

    d. $\left(\dfrac{5t}{5t}\right)^3$ _____

    h. $\left(\dfrac{x^3 y^2}{x^2 y^0}\right)^2$ _____

Negotiables (You must do at least two of these.)

1. Simplify the expressions using properties of exponents. Reduce the fractions to lowest terms. Your answers should have no negative exponents.

    a. $\dfrac{x^2 y^3}{xy^2}$ _____

    c. $\dfrac{z^5 y^7}{z^9 y^4}$ _____

    b. $\dfrac{x^5 t^3}{x^6 t^4}$ _____

    d. $\dfrac{x^5 y^7 z^4}{x^9 y^4 z^6}$ _____

2. Find the missing exponent(s) or variables to make the equation true.

a. $\dfrac{x^{\square}}{x^2} = x^3$ _____

c. $\dfrac{t^{-6}}{t^{\square}} = t^2$ _____

b. $\dfrac{x^7}{x^{\square}} = x^5$ _____

d. $\dfrac{x^{\square}}{x} = x^5$ _____

3. Answer the questions using complete sentences.

a. Is $\dfrac{y^2}{z^{-3}}$ simplified? Explain. If not, simplify it.

b. Is $\dfrac{y^2}{-z^3}$ simplified? Explain. If not, simplify it.

Options (You may choose to do these problems.)

Simplify the expressions. The answers should have no negative exponents.

1. $\dfrac{12x^5 y^{-3}}{3x^6 y^4} \cdot \left(\dfrac{-3x^3 y^{-2}}{x^5 y^3} \right)^3$

2. $\left(\dfrac{-6x^{-6} y^5}{2x^5 y^{-2}} \right)^2 \cdot \left(\dfrac{-x^4 y^2}{x^{-2} y^{-3}} \right)^{-3}$

Lesson 6: Division Properties of Exponents

Hexagon Puzzle

Directions: Solve any problem in a hexagon in the far left-hand column. If you get the answer correct, choose a problem from the next column that is adjacent to the first problem. If you get a problem wrong, move to another problem in the same column adjacent to the hexagon and solve it. Make a path to the right by solving the problems from left to right. You must complete a continuous path of problems from left to right.

A→Simplify the expression.
$$\frac{a^6}{a^2}$$

B→Simplify the expression.
$$\frac{a^9}{a^5}$$

C→Simplify the expression.
$$\frac{5^6}{5^5}$$

D→Simplify the expression.
$$\frac{t^6}{t^4}$$

E→Simplify the expression.
$$\frac{a^6}{a^9}$$

F→Simplify the expression.
$$\frac{4^2}{4^5}$$

G→Simplify the expression.
$$4^{-2}$$

H→Simplify the expression.
$$y^{-5}$$

I→Simplify the expression.
$$x^2 \cdot x^0$$

J→Simplify the expression.
$$\left(\frac{a^6}{a^2}\right)^0$$

K→Simplify the expression.
$$\left(\frac{a^3}{a^2}\right)^{-2}$$

L→Simplify the expression.
$$\left(\frac{a^{-4}}{a^2}\right)^2$$

M→Simplify the expression.
$$(x^2)^{-2}$$

N→Simplify the expression.
$$(x^2 \cdot y^{-4})^{-2}$$

O→Simplify the expression.
$$(2t^{-3} \cdot x^3)^{-2}$$

P→Simplify the expression.
$$(-2t^2 \cdot w^2)^{-1}$$

Q→Simplify the expression.
$$\frac{2x^9}{-4x^5}$$

R→Simplify the expression.
$$\frac{4x^5}{-2x^{-5}}$$

S→Simplify the expression.
$$\frac{3t^{-3}}{t^5}$$

T→Simplify the expression.
$$\frac{-4t^2}{8t^5}$$

U→Find the missing exponent.
$$\frac{x^\square}{x} = x^3$$

V→Find the missing exponent.
$$\frac{x^5}{x^\square} = x^2$$

W→Find the value of x.
$$5^{-2} = \frac{1}{x}$$

X→Find the missing exponent.
$$\frac{x^6}{y^\square} = \left(\frac{x^2}{y}\right)^3$$

Exponent Rules and Exponential Functions

Differentiating Instruction in Algebra 1 © Prufrock Press Inc. • Permission is granted to photocopy or reproduce this page for single classroom use only.

Lesson 6: Division Properties of Exponents

Hexagon Puzzle

Directions: Solve any problem in a hexagon in the far left-hand column. If you get the answer correct, choose a problem from the next column that is adjacent to the first problem. If you get a problem wrong, move to another problem in the same column adjacent to the hexagon and solve it. Make a path to the right by solving the problems. You must complete a continuous path of problems from left to right.

A→Simplify the expression.
$$\frac{a^6}{a^9}$$

B→Simplify the expression.
$$\frac{4^2}{4^5}$$

C→Simplify the expression.
$$4^{-2}$$

D→Simplify the expression.
$$y^{-5}$$

E→Simplify the expression.
$$\left(\frac{a^6}{a^2}\right)^0$$

F→Simplify the expression.
$$\left(\frac{a^3}{a^2}\right)^{-2}$$

G→Simplify the expression.
$$\left(\frac{a^{-4}}{a^2}\right)^2$$

H→Simplify the expression.
$$\left(\frac{a^{-4}}{a^2}\right)^{-2}$$

I→Simplify the expression.
$$(x^2 \cdot y^{-4})^{-2}$$

J→Simplify the expression.
$$(2t^{-3} \cdot x^3)^{-2}$$

K→Simplify the expression.
$$(-2t^2 \cdot w^2)^{-1}$$

L→Simplify the expression.
$$(2t^3 \cdot w^{-2} \cdot z)^{-1}$$

M→Simplify the expression.
$$\frac{2x^9}{-4x^5}$$

N→Simplify the expression.
$$\frac{4x^5}{-2x^{-5}}$$

O→Simplify the expression.
$$\frac{3t^{-3}}{t^5}$$

P→Simplify the expression.
$$\frac{-4t^2}{8t^5}$$

Q→Find the missing exponent.
$$\frac{x^{\square}}{x} = x^3$$

R→Find the missing exponent.
$$\frac{x^{-5}}{x^{\square}} = x^2$$

S→Find the value of x.
$$5^{-2} = \frac{1}{x}$$

T→Simplify the expression.
$$\frac{x^6}{y^{\square}} = \left(\frac{x^2}{y}\right)^3$$

U→Find the missing value.
$$\frac{\square}{x^3} = x^{-3}$$

V→Find the missing exponent.
$$\frac{z^9}{s^5} = \left(\frac{s^5}{z}\right)^{\square}$$

W→Find the missing exponents.
$$\frac{t^6}{w^{\square}} = \left(\frac{t^2}{w^3}\right)^{\square}$$

X→Find the missing exponent.
$$\frac{x^4}{y^2} = \left(\frac{x^2}{y}\right)^{\square}$$

Exponent Rules and Exponential Functions

★

Lesson 6: Division Properties of Exponents

Hexagon Puzzle

Directions: Solve any problem in a hexagon in the far left-hand column. If you get the answer correct, choose a problem from the next column that is adjacent to the first problem. If you get a problem wrong, move to another problem in the same column adjacent to the hexagon and solve it. Make a path to the right by solving the problems from left to right. You must complete a continuous path of problems from left to right.

A→Simplify the expression.
$$\left(\frac{a^6}{a^2}\right)^0$$

B→Simplify the expression.
$$\left(\frac{a^3}{a^2}\right)^{-2}$$

C→Simplify the expression.
$$\left(\frac{a^{-4}}{a^2}\right)^2$$

D→Simplify the expression.
$$\left(\frac{a^{-4}}{a^2}\right)^{-2}$$

E→Simplify the expression.
$$(x^2 \cdot y^{-4})^{-2}$$

F→Simplify the expression.
$$(2t^{-3} \cdot x^3)^{-2}$$

G→Simplify the expression.
$$(-2t^2 \cdot w^2)^{-1}$$

H→Simplify the expression.
$$(2t^3 \cdot w^{-2} \cdot z)^{-1}$$

I→Find the missing exponent.
$$\frac{x^\Box}{x} = x^3$$

J→Find the missing exponent.
$$\frac{x^{-8}}{x^\Box} = x^2$$

K→Find the value of x.
$$5^{-2} = \frac{1}{x}$$

L→Simplify the expression.
$$\frac{x^6}{y^\Box} = \left(\frac{x^2}{y}\right)^3$$

M→Find the missing value.
$$\frac{\Box}{x^3} = x^{-3}$$

N→Find the missing exponent.
$$\frac{z^9}{s^5} = \left(\frac{s^5}{z}\right)^\Box$$

O→Find the missing exponents.
$$\frac{t^6}{w^\Box} = \left(\frac{t^2}{w}\right)^3$$

P→Find the missing exponent.
$$\frac{x^4}{y^2} = \left(\frac{x^2}{y}\right)^\Box$$

Q→Find the missing exponents.
$$\left(\frac{s^4}{z^\Box}\right)^{-1} = \frac{z^3}{s^\Box}$$

R→Find the missing exponents.
$$\left(\frac{x^\Box}{y^3}\right)^2 = \frac{x^4}{y^\Box}$$

S→Simplify.
$$\frac{x^2 \cdot x^3}{z^{-5}}^{-2} = \frac{1}{z^\Box}$$

T→Find the missing exponent and value.
$$(6x^3 \cdot x^{-5} \cdot x^6)^\Box = \Box x^8$$

U→Find the missing exponent.
$$\left(\frac{t^2 z^5}{t}\right)^3 = t^{21} z^{15}$$

V→Find the missing exponents.
$$\left(\frac{x^2 y^3}{x^\Box y^\Box}\right)^{-2} = \frac{1}{x^8 y^{10}}$$

W→Find the missing exponent.
$$\left(\frac{x^{-5} y^{-3}}{x^4 y^3}\right)^\Box = x^9 y^6$$

X→Find the missing exponent.
$$\left(\frac{x^\Box y^4}{x^{-2} z^5}\right)^2 = \frac{x^{10} y^8}{z^{10}}$$

Exponent Rules and Exponential Functions

Lesson 6: Division Properties of Exponents
Exit Slip

1. Explain in words the following division properties of exponents:

    a. $\dfrac{a^m}{a^n}$

    b. $\left(\dfrac{a}{b}\right)^m$

2. Create two new problems for each of the above properties, and solve the problems.

Lesson 7: Wrap Up and Assessment
Create Your Own Children's Book Lesson Plan

Introduction
Students will write a children's book to demonstrate their understanding of exponential growth or decay as a summative assessment for the end of this unit. This project is easily differentiated by reducing the number of pages or providing a story and asking a student to identify the exponential and then rewriting the ending of the story to show understanding. All students will be required to peer review each other's work.

Materials Needed
- Children's book as an example
- Two or more sheets of plain paper per student (folded in half)
- Colored paper for covers
- Colored pencils, crayons, markers
- Stapler

Lesson
- Discuss *One Grain of Rice* (or some other's children book) or the "Doubled Pennies" problem that models exponential models. Discuss what makes these stories about exponential functions.
- Talk about what makes a good children's book (e.g., vivid illustrations, interesting characters, intriguing story line).
- Provide students with the Create Your Own Children's Book Project Requirements handout that includes the rubric outlining how the project will be assessed. Discuss the project components. Talk about the components of an effective peer review and assign peer review partners.

Summary/Reflection
Students should share their books with the class or in small groups. Have students give constructive feedback to each author. This could be done before the books are turned in for final grading to give students an opportunity to make changes to their stories based on student recommendations. Questions to consider include:
- Did students demonstrate an understanding of exponential functions?
- Was enough time provided to complete the project?
- Was the project enjoyable for the majority of students?
- What would you do differently next time?

Lesson 7: Wrap Up and Assessment

Create Your Own Children's Book Project Requirements

Goal

You will create your own children's book that demonstrates your understanding of exponential growth or decay.

Project Requirements

Following are the project requirements, which are also listed in the project rubric.

➤ You must have a cover for your book that includes the title, picture of your main character, author, and date.

➤ The cover must be glued or stapled to the inside pages of the book.

➤ Your book must have at least eight pages and must have illustrations on each page.

➤ Your book must include at least the first eight terms in the sequence of your exponential growth or decay model.

➤ Your book must be reviewed by a peer, who will check for spelling and verify that you have properly modeled an exponential situation.

➤ The back cover of the book must contain your equation that models your exponential situation.

Create Your Own Children's Book Rubric

Criteria	Excellent (2 points)	Acceptable (1 point)	Poor (0 points)	Points Earned
Front and back cover elements	All elements are included.	One element is missing.	More than one element is missing.	
Book has a minimum of eight pages	At least eight pages are included.	One page is missing.	More than one page is missing.	
Illustrations on every page	Illustrations are appropriate and creative.	Illustrations are missing on one page or lack creativity.	Illustrations are missing on more than one page and lack creativity.	
First eight terms of exponential growth or decay model included	Book includes first eight terms.	Eight terms are included but one is incorrect.	Fewer than eight terms are included or more than one is incorrect.	
Peer reviewed	The book was peer reviewed.		The book was not peer reviewed.	
Neatness	The final product is neat and easy to read.	Most of the product is neat and easy to read.	The final product is not neat, nor is it easy to understand.	
			Total: _____/12= _____%	

Comments:

129

Exponent Rules and Exponential Functions

Lesson 7: WRap Up and Assessment

Exit Slip

1. Discuss three significant things that you learned about exponential functions in this unit.

2. Which areas of exponential functions do you still find challenging?

Exponent Rules and Exponential Functions

Unit 4

QuadRatic Functions

The final unit focuses on quadratic functions. Students will continue their study of nonlinear functions by identifying quadratic functions in tables, graphs, and equations and identifying the characteristics of quadratics. Real-life applications introduced through story problems are included in most lessons. The final project, which could be used as an authentic assessment, is a catapult project.

This unit begins with a pretest and four real-life applications of quadratic functions that can be discussed in small groups and then as a larger group. Many of the activities will offer the students an opportunity to choose learning activities according to their learning style, personal interests, and readiness level.

What Do We Want Students to Know?

Common Core State Standards Addressed:	Big Ideas
• A.SSE.3a, 3b • A.APR.3 • A.CED.2 • A.REI.4a, 4b • F.IF.4, 6, 7a, 8a, 9 • F.BF.1	• Quadratics are parabolas and must have a 2 as their highest exponent. • The zeros/solutions are where the parabola crosses the x-axis. • There are different ways to solve quadratics. • Different forms of quadratics provide different key information. • Quadratics can model many real-world situations.
	Essential Questions • What does it mean to find the zeros/solutions? • What is a parabola? • How do I determine the best method to solve a quadratic? • What information do I need to sketch a quadratic?

Critical Vocabulary

Parabola

y-intercept

Maximum value

Quadratic function

Second differences

Function

Zeros (roots)

Minimum value

Quadratic formula

Terms

x-intercept

Axis of symmetry

Vertex

First differences

Unit Objectives

As a result of this unit, students will know:

➤ quadratic functions have an independent variable raised to a second power,

➤ when quadratic functions are graphed they are shaped like parabolas, and

➤ all quadratic functions have a vertex, a line or symmetry, and a minimum or maximum.

As a result of this unit, students will understand that:

➤ quadratic functions can be written in three different forms, and each form provides key information about the quadratic;

➤ quadratic functions can have zero, one, or two solutions;

➤ quadratic functions can be solved using various methods; and

➤ quadratic functions model real-life situations.

As a result of this unit, students will be able to:

➤ identify a quadratic relationship from a graph, table, or equation;

➤ find the vertex, axis of symmetry, and roots (zeros) from a graph, table, or equation;

➤ solve quadratic equations by graphing, factoring, completing the square, or using the quadratic formula; and

➤ interpret maximum and minimum points and intercepts in real-life applications.

Launch Scenarios

➤ After a soccer game, teammates shake hands to congratulate each other. How many handshakes are exchanged between a team with seven players? A team of nine players? A team of 11 players? (Lesson 2)

➤ You are putting a fence in your backyard to surround your garden, but you are limited to 24 meters of fencing. How should you arrange the fencing in order to maximize the amount of garden space? (Lesson 2)

➤ You want to put a decorative stone walking path along two adjacent sides of a water garden. The water garden measures 4 ft by 10 ft. How wide should the path be if there is enough decorative stone to cover 51 square feet? (Lesson 4)

➤ You are building a catapult to launch objects in your algebra class. How many feet away from the landing site do you have to place your catapult in order to hit the target? (Lesson 6)

Unit Overview: Quadratic Functions

(Assumes a 50–60-minute class period)

Lesson	Whole-Class/Small-Group Discussion/Activities	Individualized Learning Activities
Preassessment (1/2 period)	Give the preassessment at least several days prior to beginning the unit (20 minutes)	
Lesson 1: Introduction (2 periods)	Share results of the preassessment (20 minutes) Introduce launch scenarios and discuss possible methods of solving (10 minutes)	
		Learning Targets and Study Guide (15 minutes) Vocabulary Choice Board (15 minutes)
	Introduce checkerboard activity (20 minutes)	
		Exit Slip (10–15 minutes)

Lesson	Whole-Class/Small-Group Discussion/Activities	Individualized Learning Activities
Lesson 2: Characteristics of Quadratics (3 periods)	Hold a class discussion on characteristics of quadratics (25 minutes) The Handshake Challenge (20 minutes) Hold a group discussion on launch scenarios (15 minutes)	
		Agenda (30 minutes) Tic-Tac-Toe Board (30 minutes)
	Forms of Quadratic Equations (20 minutes)	
		Hexagon Puzzles (30 minutes) Exit Slip (10–15 minutes)
Lesson 3: Solving Using the Quadratic Formula (2 periods)	Hold a class discussion on solving using quadratic formula (20 minutes)	
		Agenda (30 minutes) Think Dots (40 minutes) Exit Slip (10–15 minutes)
Lesson 4: Solving by Factoring (2.5 periods)	Checkerboard Activity (20 minutes) Hold a class discussion on solving by factoring (20 minutes)	
		Agenda $x^2 + bx + c$ (30 minutes) Agenda $ax^2 + bx + c$ (30 minutes) Tic-Tac-Toe Board (30 minutes) Exit Slip (10–15 minutes)

Lesson	Whole-Class/Small-Group Discussion/Activities	Individualized Learning Activities
Lesson 5: Solving by Completing the Square (2 periods)	Hold a class discussion on solving by completing the square (20 minutes)	
		Agenda (40 minutes) Think Dots (45 minutes) Exit Slip (10–15 minutes)
Lesson 6: Wrap Up and Assessment (3–4 periods)	Collect learning targets and study guide and vocabulary choice board projects	
		TriMind Activity (30 minutes) RAFT Activity (40 minutes)
	Catapult Activity (2–3 periods)	
		Exit Slip (10–15 minutes)
Lesson 7: Who Uses Mathematics?	Structured Academic Controversy (50 minutes)	

Name: _____ Hour/Block: _____ Date: _____

QuadRatic Functions

Preassessment

Directions: Solve the following problems to the best of your ability, and then rate your confidence in your answer in the space provided. Skip questions you do not know. This will not be graded.

1. Graph $y = x^2 + 2x + 1$. Identify the vertex and axis of symmetry, and identify if the quadratic has a maximum or minimum.

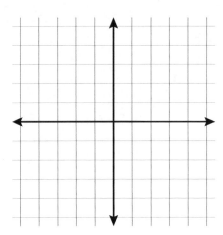

_____ I'm sure

_____ I'm not sure

2. Do the following tables of data represent quadratic functions?

a.

x	−3	−2	−1	0	1	2	3
y	26	15	8	5	6	11	20

b.

x	−3	−2	−1	0	1	2	3
y	−11	−9	−7	−5	−3	−1	1

_____ I'm sure

_____ I'm not sure

3. Solve $y = 6x^2 - x - 2$ using factoring.

_____ I'm sure

_____ I'm not sure

4. Solve $-2x^2 + 3x + 4$ using the quadratic formula.

_____ I'm sure

_____ I'm not sure

5. Solve $x^2 + 6x + 5$ by completing the square.

_____ I'm sure

_____ I'm not sure

6. Write a quadratic function that satisfies each of the conditions below.

   a. A quadratic function with two solutions _____

   b. A quadratic function with one solution _____

   c. A quadratic function with zero solutions _____

   _____ I'm sure

   _____ I'm not sure

7. The height of a baseball thrown into the air can be represented by the function $h(t) = -16t^2 + 70t + 5$ where $h(t)$ is the height of the ball in feet after t seconds. Answer the questions below.

   a. At what height is the ball when it is thrown according to the equation? _____

   b. Find $h(0)$ and give a real-world meaning for this value. _____

   c. How high is the ball when $t = 3$ seconds? _____

   d. When is the ball 30 feet above the ground? Does this happen more than once? _____

   e. What is the highest the ball travels? _____

   f. When does the ball hit the ground? _____

   _____ I'm sure

   _____ I'm not sure

QuadRatic Functions

Learning Targets and Study Guide

At the end of this unit, every student should be able to say:

Learning Target	Explain and Give an Example
I can determine if a table of data represents a quadratic relationship. Initial here when mastered _____	
I can determine if an equation represents a quadratic relationship. Initial here when mastered _____	
I can draw graphs of quadratics with zero, one, and two solutions. Initial here when mastered _____	
I can solve a quadratic function by factoring. Initial here when mastered _____	
I can solve a quadratic function by graphing. Initial here when mastered _____	
I can solve a quadratic function by completing the square. Initial here when mastered _____	
I can solve a quadratic function by using the quadratic formula. Initial here when mastered _____	
I can solve a real-life story problem involving a quadratic function and identify the meaning of the y-intercept, the zeros, and the vertex. Initial here when mastered _____	

Quadratic Functions

Quadratic Functions

Vocabulary Choice Board

Directions: Using the critical vocabulary for the unit, choose one of the following methods to demonstrate your knowledge of the mathematical terms.

LINCS Cards	Graphic Organizer	News Broadcast
Prepare a LINCS card for each vocabulary word to demonstrate your understanding of the vocabulary words. *(logical–mathematical)*	Construct a graphic organizer of your choice to define each word and demonstrate your understanding of the vocabulary words and how they fit into the current unit. *(verbal/language–mathematical)*	Write a news broadcast that defines all of the vocabulary words, and provide stories that demonstrate your understanding of the vocabulary words. Turn in your news broadcast, and then present it to the class. *(intrapersonal–language)*
Nature Survival	**Creative Story**	**Poster Board**
Prepare a nature survival guide about how each of the vocabulary words would be used if you were stranded on a deserted island. The guide must demonstrate your understanding of the vocabulary words. *(natural–language)*	Write a creative story with color illustrations that demonstrates your understanding of the vocabulary words. *(verbal–language)*	Prepare an 11" x 14" poster board that contains the definitions of the vocabulary words and includes one real-life example for each word that demonstrates your understanding of the definition. *(language–spatial)*
Cartoon	**Rap or Song**	**Acting Out**
Prepare a cartoon strip with at least five frames that defines each word and is illustrated with appropriate pictures to demonstrate your understanding of the vocabulary words. *(visual–spatial)*	Write a rap or a song containing all of the vocabulary words and their correct definitions. Perform your rap or song for the class. *(musical–language)*	Write a script and then act out the words with their correct definitions. Use props if appropriate. *(kinesthetic–language)*

Critical Vocabulary

Parabola	Function	x-intercept
y-intercept	Zeros (roots)	Axis of symmetry
Maximum value	Minimum value	Vertex
Quadratic function	Quadratic formula	First differences
Second differences	Terms	

QuadRatic Functions
Vocabulary Choice Board Rubric

Grading scale:	3 Well done	2 Close to requirements		1 Needs more work
Choice	**Definitions**	**Illustrations and/ or creativity**	**Neatness**	**Other**
LINCS cards	Definitions demonstrate knowledge of vocabulary. Score _____	The illustrations are appropriate and creative. Score _____	The cards are easy to read. Score _____	Semantic connections between words are demonstrated. Score _____
Graphic organizer	Definitions demonstrate knowledge of vocabulary. Score _____	The graphic organizer creatively displays the definitions. Score _____	The graphic organizer is easy to read. Score _____	The graphic organizer choice is appropriate for this assignment. Score _____
News broadcast	Definitions demonstrate knowledge of vocabulary. Score _____	The news broadcast is creative and holds the attention of the audience. Score _____	The news broadcast is easy to read. Score _____	Good presentation skills are demonstrated. Score _____
Nature survival guide	Definitions demonstrate knowledge of vocabulary. Score _____	The guide contains illustrations that are appropriate. Score _____	The guide is easy to read. Score _____	The final product resembles a survival guide. Score _____
Creative story	Definitions demonstrate knowledge of vocabulary. Score _____	The story contains illustrations that are appropriate. Score _____	The story is easy to read. Score _____	The story is interesting. Score _____
Poster board	Definitions demonstrate knowledge of vocabulary. Score _____	The poster board includes illustrations that are appropriate. Score _____	The poster board is easy to read. Score _____	Real-life examples are included for each vocabulary term. Score _____
Cartoon	Definitions demonstrate knowledge of vocabulary. Score _____	The cartoon includes illustrations that are appropriate. Score _____	The cartoon is easy to read. Score _____	The cartoon contains a minimum of five frames. Score _____
Rap or song	Definitions demonstrate knowledge of vocabulary. Score _____	The rap or song is creative. Score _____	The lyrics are easy to read. Score _____	Good performance skills are demonstrated. Score _____
Acting out	Definitions demonstrate knowledge of vocabulary. Score _____	The script is creative. Score _____	The script is easy to read. Score _____	Good performance skills are demonstrated. Score _____
				Total: _____/12= _____%

Comments:

Quadratic Functions

Lesson 1: Introduction

Checkerboard Small-Group Activity Lesson Plan

Purpose: Students will work in small groups (in pairs or groups of three) to put together the checkerboard puzzle that reviews identifying solutions to linear equations and identifying the slope and *y*-intercept from an equation.

Prerequisite Knowledge	Materials Needed
Students should know: • how to list factors of numbers, • the exponent rules for multiplication, and • how to find greatest common factors for numbers and variables.	• One copy of the checkerboard cut into squares for each small group of students • Scrap paper for solving equations

Lesson

> ➤ Teachers may want to use use random grouping. Each group should receive a packet of the puzzle pieces that are in no particular order.

> ➤ Students are asked to put together the checkerboard by matching pieces that have questions and answers that go together. Their finished puzzle is the message "DOGS RULE CATS LOVE MICE."

> ➤ Students may notice that there are puzzle pieces that are blank on one side. They may figure out that these pieces belong on the left or right border of the puzzle.

Discussion Questions

> ➤ What message is revealed when the puzzle is completed?
> ➤ How do you find the greatest common factors of numbers/variables?
> ➤ What is the rule for multiplying exponents?
> ➤ How do you use the distributive property to eliminate parenthesis?
> ➤ What is the formula for finding the area of a square/rectangle?
> ➤ What is the formula for finding the perimeter of a square/rectangle?
> ➤ What is a prime number?

Quadratic Functions

Factors of 15	1, 4, 6, 12	12	5x + 6
D 1, 2, 3, 6	**O** Factors of 6	**G** 1, 2, 3, 6, 9, 18 · Factors of 18 · 2x + 2	**S** 2(x + 1)
Factors of 12	Product of y · y	GCF of 12 and 18	2x + 4x
1, 2, 3, 4, 6, 12	y^2	9	6x
R 7 + x + 8	**U** 2x + 5x + 8	**L** 1, 3, 9, 27 · Factors of 27	**E** $4x^2$ · $(2x)^2$
Area of a square with sides of 5x	$5(x^2 + x)$	Multiples of 3	GCF of $2x^2$ and 4x
$25x^2$	$5x^2 + 5x$	3, 6, 9, 12, 15	2x
C −3y	**A** 5y − 8y	**T** 27 · GCF of 27 and 81	**S** $6x^2 + 3xy$ · 3x(2x + y)
GCF of 5x and 30xy	Area of a square with sides of 7y	3(2x + 4y)	Perimeter of a rectangle with sides of x and z
5x	$49y^2$	6x + 12y	2x + 2z
L 1	**O** GCF of 29 and 35	**V** 5z + 30x · 5(z + 6x)	**E** 1, 3, 7, 21 · Factors of 21
Multiples of 6	$5(x^2 − 5)$	GCF of $3xy^2$ and 9xy	GCF of 14 and 56
6, 12, 18, 24, 30	$5x^2 − 25$	3xy	14
M 15x	**I** GCF of 15x and $45x^2$	**C** Perimeter of a rectangle with sides of x and z · x + x + z + z	**E** $(s)^2$ · Area of a square
6x + 2y	$5x^2 − 20$	5(z + 2x)	7

Lesson 1: Introduction

Exit Slip

1. Explain how to complete the table of values using the equation $y = x^2 + 2x - 1$.

x	y
-3	
-2	
-1	
0	
1	

2. Sketch the graph with the five points from Question 1 and describe the shape of the graph.

   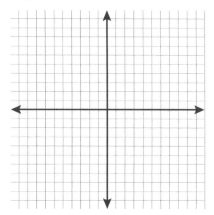

3. What is the coordinate of the lowest point on the graph?

4. For your sketch in Question 2 above, draw the vertical line that would divide the graph in half. Write the equation for the axis of symmetry.

Quadratic Functions

Lesson 2: Characteristics of Quadratics
The Handshake Challenge

Challenge: After a soccer game, teammates shake hands to congratulate each other. How many handshakes are exchanged between a team with seven players? A team of nine players? A team of 11 players?

1. Working in your group, determine two ways to solve this challenge. Be prepared to discuss your two methods with the class.

2. With your group, complete the following table for the number of people in your group by modeling the challenge. Shake hands with everyone in the group just once and record the number of handshakes.

# of people (x)	2	3	4	5	6	7	8	9	n
# of handshakes (y)									

3. Combine your group with another group and continue to complete the table. Do you notice any patterns? What would be the number of handshakes for 10 people? For 11 people?

4. Working with your original group, write an equation to represent these data.

Extension Questions

1. There are 12 lacrosse teams at a tournament. How many matches must be played in order for each of the teams to play every other team one time?

2. Triangular numbers are numbers represented by the shapes below. How many dots will be in the fourth picture? The fifth picture? The nth picture?

Lesson 2: Characteristics of Quadratics
Agenda

Imperatives (You must do all three of these.)

1. Identify the a, b, and c in the following equations in standard form.

   a. $y = x^2 + 2x - 5$ $a =$ _____ ; $b =$ _____ ; $c =$ _____

   b. $y = -2x^2 - 4x + 6$ $a =$ _____ ; $b =$ _____ ; $c =$ _____

   c. $y = -x^2 + 2x$ $a =$ _____ ; $b =$ _____ ; $c =$ _____

   d. $y = 5x^2 + 2$ $a =$ _____ ; $b =$ _____ ; $c =$ _____

2. Find the axis of symmetry for the following quadratic functions using the formula $x = \dfrac{-b}{2a}$.

   a. $y = 2x^2 - 4x - 5$

   b. $y = -2x^2 + 4x + 5$

   c. $y = -x^2 + 6x + 1$

   d. $y = 2x^2 - 10x + 2$

3. Using the formula $x = \dfrac{-b}{2a}$, find the (x, y) coordinate of the vertex for the following quadratic functions. Find the x-value, and then substitute it into the equation and solve for y.

   a. $y = 2x^2 - 4x + 5$

   b. $y = x^2 - 2x + 3$

   c. $y = 3x^2 - 12x + 16$

Negotiables (You must do at least two of these.)

For each of the following quadratic functions, determine the vertex and axis of symmetry using a graphing calculator or the formula $x = \dfrac{-b}{2a}$. Also identify whether the vertex is a maximum value or a minimum value.

1. $y = x^2 - 2$

   Vertex: _____

   Line of symmetry: _____

   Minimum/maximum: _____

Quadratic Functions

2. $y = 2x^2 + 4x + 1$

   Vertex:_____

   Line of symmetry: _____

   Minimum/maximum:_____

3. $y = -x^2 - x - 6$

   Vertex:_____

   Line of symmetry: _____

   Minimum/maximum:_____

4. $y = (x-1)^2 - 4$

   Vertex:_____

   Line of symmetry: _____

   Minimum/maximum:_____

5. $y = -(x+2)^2 + 4$

   Vertex:_____

   Line of symmetry: _____

   Minimum/maximum:_____

6. $9 = x^2 - y$

   Vertex:_____

   Line of symmetry: _____

   Minimum/maximum:_____

Options (You may choose to do this problem.)

1. When an object is thrown, it can be modeled with the equation $h(t) = -16t^2 + vt + s$ where h = height (feet), t = time in motion (seconds), s = initial height (feet), and v = initial velocity (feet per second). A batter hitting a pitched baseball can be modeled by $h(t) = -16t^2 + 70t + 2.5$ where h = height (feet) and t = time in motion (seconds).

   a. Using the equation, what is the initial velocity of the baseball? _____ft/sec

   b. What is the initial height of the baseball? _____ft

   c. How high is the baseball after 3.5 seconds? _____ft

   d. How long does it take for the baseball to hit the ground? _____sec

Quadratic Functions

Lesson 2: Characteristics of Quadratics
Tic-Tac-Toe Board

Directions: Choose three problems in a row, in a column, or on a diagonal. Solve the problems and show your work on a separate sheet of paper. Circle the problems you choose on this page and then reference them on your work paper with the number from the square.

1	**2**	**3**
Identify whether the following equations are linear, exponential, or quadratic. a. $y = 2x - 7$ b. $y = x^2 - 5x$ c. $2y = 7 + 4x$ d. $y = 2^x - 5$ e. $x^2 = 2y + 3$	Determine whether the following quadratic functions open upward or downward. Circle the coefficient that determines the direction of the parabola. a. $y = x^2 + 2x - 4$ b. $y = -5x^2 - 9x + 5$ c. $y = -8x + 3x^2 - 2$ d. $-2y = -3x^2 + x - 1$	Create quadratic equations that meet the following criteria: a. Opens upward b. Opens downward c. Has no x-intercepts d. Has two x-intercepts
4	**5**	**6**
Prove that the following table of data represents a quadratic function. 	x	y
---	---
-2	20
-1	9
0	2
1	-1
2	0		Find the vertex and axis of symmetry for the following quadratics (hint: find a, b, c). a. $y = x^2 + 3x - 2$ b. $y = -2x^2 - 3x$ c. $2y = 4x^2 + 6x - 1$	Find the x-intercepts (zeros) of the following quadratic functions (write your answers as coordinate pairs). a. $y = x^2 + x - 2$ b. $y = x^2 + x - 6$
7	**8**	**9**
Sketch graphs of the following quadratics and then describe the differences in the shape, the vertex, and the direction. a. $y = x^2 + 2$ b. $y = 5x^2 + 6$ c. $y = -x^2 + 2$	Find the x-intercepts (zeros) of the following quadratic functions (write your answers as coordinate pairs). a. $y = -x^2 + 2x$ b. $y = x^2 - 2x + 1$	Fill in the missing value in the table, which represents a quadratic function. 	x	y
---	---
-3	13
-2	6
-1	3
0	
1	9
2	18	

Quadratic Functions

Lesson 2: Characteristics of Quadratics

Forms of Quadratic Equations

General/Standard Form	Vertex Form	Factored Form
$y = ax^2 + bx + c$, where c is the starting value (y-intercept) of the function	$y = a(x - h)^2 + k$, where (h, k) is the vertex of the parabola	$y = a(x - r_1)(x - r_2)$, where r_1 and r_2 are the zeros of the quadratic function

Directions: Identify the form for each of the following quadratic equations. If the function is written in general/standard form, identify the starting value. For all equations, find the vertex (as a coordinate pair), write the equation for the axis of symmetry (in the form $x =$), and then find the zeros. There is one sample problem. Round decimals to two places if necessary.

Equation	Form	Starting Value	Vertex	Axis of Symmetry	Zeros
$y = 2(x + 2)^2 - 4$	vertex form	—	$(-2, -4)$	$x = -2$	$-3.41, -0.59$
$y = 2(x - 2)^2 + 5$					
$y = 3(x - 2)(x - 4)$					
$y = -6(x - 3)^2 + 6$					
$y = 2x^2 + 4x - 6$					
$y = x^2 - 3x + 7$					

When you have completed the problems, check your answers. Each box is worth 1 point.

My score is _____.

If you scored 0–17 points: You need additional practice and will complete the hexagon puzzle labeled with a moon.

If you scored 18–22 points: You are working on level but need a little more practice. You will complete the hexagon puzzle labeled with a sun.

If you scored 23–25 points: You are ready to be challenged and will complete the hexagon puzzle labeled with a star.

Quadratic Functions

Lesson 2: Characteristics of Quadratics

Hexagon Puzzle

Directions: Solve any problem in a hexagon in the far left-hand column. If you get the answer correct, choose a problem from the next column that is adjacent to the first problem. If you get a problem wrong, move to another problem in the same column adjacent to the hexagon and solve it. Make a path to the right by solving the problems. You must complete a continuous path of problems from left to right.

A→Identify the starting value for the quadratic.
$y = x^2 - 2x + 3$

B→Identify the vertex for the quadratic.
$y = (x-2)^2 - 3$

C→Identify the starting value for the quadratic.
$y = x^2 - 2x - 8$

D→Identify the zeros or roots for the quadratic.
$y = (x-2)(x+7)$

E→Identify the zeros or roots for the quadratic.
$y = (x-1)(x+4)$

F→Change the equation $y = 2x^2 - 2x + 3$ so that it opens downward.

G→Identify the vertex for the quadratic.
$y = (x+3)^2 + 1$

H→Change the equation $y = 2(x+3)^2 + 1$ so the vertex is $(3, -1)$.

I→What is another word for the starting value in the general form of a quadratic?

J→Name the characteristic of a parabola that can be found in $y = (x+1)^2 - 2$.

K→Name the characteristic of a parabola that can be found in $y = (x+9)(x-5)$.

L→Name the characteristic of a parabola that can be found in $y = x^2 - 2x + 1$.

M→Identify the axis of symmetry for the function $y = (x-2)^2 - 1$.

N→Identify the axis of symmetry for the function $y = x^2 + 2x - 3$.

O→Identify the axis of symmetry for the function $y = (x+2)^2 + 9$.

P→Identify the axis of symmetry for the function $y = 2x^2 - 4x + 9$.

Q→Identify the zeros or roots of the function $y = (x-1)(x+4)$.

R→Identify the zeros or roots of the function $y = (x+5)(x+2)$.

S→Identify the zeros or roots of the function $y = x^2 - 2x - 3$.

T→Identify the zeros or roots of the function $y = 2x^2 + 8x + 6$.

U→Create a quadratic function that has $(1,3)$ as the vertex.

V→Create a quadratic function that has zeros of 3 and 6.

W→Create a quadratic function that has $(-3,-2)$ as the vertex.

X→Create a quadratic function that has zeros of −2 and −9.

Quadratic Functions

Lesson 2: Characteristics of Quadratics

Hexagon Puzzle

Directions: Solve any problem in a hexagon in the far left-hand column. If you get the answer correct, choose a problem from the next column that is adjacent to the first problem. If you get a problem wrong, move to another problem in the same column adjacent to the hexagon and solve it. Make a path to the right by solving the problems. You must complete a continuous path of problems from left to right.

A→Identify the starting value for the quadratic. $y = 3x^2 - 2x - 1$

B→Identify the vertex for the quadratic. $y = (x+5)^2 - 1$

C→Change the equation $y = x^2 + x - 7$ so that it crosses the y-axis at 6.

D→Identify the zeros or roots for the quadratic. $y = (x+3)(x-2)$

E→Name the characteristic of a parabola that can be found in $y = (x+1)^2 - 2$.

F→Change the equation $y = -2x^2 - 2x + 3$ so that it opens in the opposite direction.

G→Name the characteristic of a parabola that can be found in $y = (x+9)(x-5)$.

H→Change the equation $y = 2(x+3)^2 + 1$ so the vertex is $(-2,8)$.

I→Write a quadratic function that has no zeros.

J→Write a quadratic function that has a vertex at $(0,4)$.

K→Write a quadratic function that has an axis of symmetry of $x = 3$.

L→Write a quadratic function that has two zeros.

M→Identify the axis of symmetry for a quadratic that has zeros at 2 and −2.

N→Identify the axis of symmetry for a quadratic that has zeros at 0 and −6.

O→Identify the axis of symmetry for a quadratic that has a zero at 5.

P→Identify the axis of symmetry for a quadratic that has zeros at 4 and −3.

Q→Create two quadratic functions that have $(1,3)$ as the vertex.

R→Create two quadratic functions that have zeros of 3 and 6.

S→Create two quadratic functions that have $(-3,2)$ as the vertex.

T→Create two quadratic functions that have zeros of −2 and −9.

U→Rewrite the equation $y = (x+3)^2 + 1$ in standard form.

V→Rewrite the equation $y = 2(x-2)^2 + 1$ in standard form.

W→Rewrite the equation $y = 2(x-3)^2 + 2$ in standard form.

X→Rewrite the equation $y = -3(x-1)^2 + 2$ in standard form.

Quadratic Functions

Name: _____   Hour/Block: _____   Date: _____

Lesson 2: Characteristics of Quadratics

Hexagon Puzzle

Directions: Solve any problem in a hexagon in the far left-hand column. If you get the answer correct, choose a problem from the next column that is adjacent to the first problem. If you get a problem wrong, move to another problem in the same column adjacent to the hexagon and solve it. Make a path to the right by solving the problems. You must complete a continuous path of problems from left to right.

A→Create a quadratic function that has no zeros or roots.

B→Change the equation $y = -2x^2 - 2x + 3$ so that it opens in the opposite direction.

C→Change the equation $y = 2x^2 + x - 1$ so that it crosses the y-axis at 6.

D→Create a quadratic function that has a vertex at $(0, 4)$.

E→Identify the axis of symmetry for a quadratic that has zeros at 6 and −2.

F→Identify the axis of symmetry for a quadratic that has zeros at 0 and −6.

G→Identify the axis of symmetry for a quadratic that has a zero at −3.

H→Change the equation $y = 2(x+3)^2 + 1$ so the vertex is $(-7, -3)$.

I→Create two quadratic functions that have $(1, 5)$ as the vertex.

J→Create two quadratic functions that have zeros at −2 and 1.

K→Create a quadratic that has an axis of symmetry of $x = -1$.

L→Create a quadratic function that has two zeros.

M→Rewrite the equation $y = (x+8)^2 - 2$ in standard form.

N→Rewrite the equation $y = 2(x-3)^2 + 2$ in standard form.

O→Rewrite the equation $y = -3(x-1)^2 + 2$ in standard form.

P→Identify the axis of symmetry for a quadratic that has zeros at 4 and −3.

Q→Translate the graph of $y = x^2$ down 5 units.

R→Translate the graph of $y = (x-1)^2$ down 2 units.

S→Translate the graph of $y = x^2$ up 3 units.

T→Translate the graph of $y = (x+1)^2 - 3$ down 5 units.

U→Describe the transformation of $y = (x+1)^2 + 4$ from $y = x^2$.

V→Transform the equation $y = (x+1)^2 - 3$ down 2 units and to the left 2 units.

W→Describe the transformation of $y = 4x^2 - 2$ from $y = x^2$.

X→Explain why the value of $f(x) = x^2 + 3x - 2$ increases as the value of x decreases from −2 to −10.

Quadratic Functions

Lesson 2: Characteristics of Quadratics

Exit Slip

1. Given a quadratic equation in standard form, $y = ax^2 + bx + c$, explain how the value of a affects the shape of the graph.

2. Given a quadratic equation in standard form $y = ax^2 + bx + c$, explain how the value of c affects the shape of the graph.

3. Explain how to find the vertex of a parabola without a calculator. Use the equation $y = 2x^2 - 3x + 1$ if you need an example.

4. Explain how to find the axis of symmetry and the significance of this line.

Quadratic Functions

Lesson 3: Solving Using the Quadratic Formula

Agenda

$$x = \frac{-b \pm \sqrt{b^2 - 4ac}}{2a}$$

Imperatives (You must do all three of these.)

1. Explain how to use the quadratic formula to solve $7 = 2x^2 - 4x + 1$.

2. Rewrite the following equations in standard form so that you can use the quadratic formula to solve them.

    a. $-1 = 2x^2 + 3x - 5$ $0 = $ _____

    b. $5 = -2x^2 - 6x$ $0 = $ _____

    c. $2 - x = 4x^2$ $0 = $ _____

    d. $4x + 6 - 5x^2 = 3x$ $0 = $ _____

3. Use the quadratic formula to solve each equation.

    a. $0 = 4x^2 - 13x + 3$ $x = $ _____

    b. $0 = 2x^2 - 4x - 30$ $x = $ _____

    c. $0 = x^2 - 3x + 2$ $x = $ _____

    d. $0 = -3x^2 + 6x + 24$ $x = $ _____

Vertical Motion Models

The following equations can be used to model falling objects and objects that are thrown that have an initial velocity. The following variables are used in these models:

h = height (ft)

t = time (sec)

s = initial height (ft)

v = initial velocity (ft per sec)

Object that is dropped: $h(t) = -16t^2 + s$

Object that is thrown: $h(t) = -16t^2 + vt + s$

Negotiables (You must do at least one of these.)

1. While on vacation in Yellowstone National Park, you accidently drop your sunglasses from an observation deck 250 feet into the canyon below. How long will it take for the sunglasses to hit the ground? Solve this using the quadratic formula. Explain why you can eliminate one of your solutions.

2. Wayne hits a tennis ball 2.5 feet off the ground at a velocity of 60 ft/sec. How long does it take the tennis ball to hit the ground?

Options (You may choose to do this problem.)

1. Using the data in the following table and your graphing calculator, write the equation to represent this data. Explain what keys you need to press on your calculator and explain each step.

x	−4	−3	−2	−1	0	1	2	3	4
y	−10	−1	4	5	2	−5	−16	−31	−50

Lesson 3: Solving Using the Quadratic Formula

Think Dots

Directions: In your groups, each student will take turns rolling the die. Answer the corresponding question on the record sheet. You may discuss your answers with your group members before recording your answer. You must answer all of the questions.

⚀	⚁	⚂
Define vertex, minimum value, maximum value, and axis of symmetry. Sketch two parabolas, one that opens upward and one that opens downward, and label these features.	Sketch pictures of quadratics with two solutions, one solution, and zero solutions.	Change the following quadratic equation so that it has two solutions. $$y = 2x^2 + 3x + 4$$
⚃	⚄	⚅
Write the quadratic formula and then write in words how to read the formula using everyday language and the correct mathematical terms.	Identify the a, b, and c values for the following equations. a. $0 = x^2 + 2x + 5$ b. $0 = -4x^2 - 3x + 1$ c. $0 = 9 - 2x^2 + x$ d. $0 = 4 - x^2 + 5x$	Using the quadratic formula, solve $x^2 + 9x + 14 = 0$. Show how to check your solutions using the original equation and a graphing calculator.

Quadratic Functions

Lesson 3: Solving Using the Quadratic Formula

Think Dots

Directions: In your groups, each student will take turns rolling the die. Answer the corresponding question on the record sheet. You may discuss your answers with your group members before recording your answer. You must answer all of the questions.

		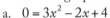
Define vertex, minimum value, maximum value, and axis of symmetry. Sketch two parabolas, one that opens upward and one that opens downward, and label these features.	Identify the a, b, and c values for the following equations. a. $0 = 3x^2 - 2x + 4$ b. $0 = -4x^2 - 3x$ c. $-2 = -4x^2 + 2x - 5$ d. $-2 = 4 - 6x^2 - x$	Using the quadratic formula, solve $x^2 + 9x + 9 = -5$. Show how to check your solutions using the original equation and a graphing calculator.
Using the quadratic formula, solve $2x^2 + 3x + 5 = -2$. Show how to check your solutions using the original equation and a graphing calculator.	A falcon diving from the top of a building to catch its prey can be modeled with the function $h(t) = -16t^2 - 200t + 100$, where $h(t)$ represents the height in feet at t seconds. a. According to the equation, how fast is the falcon traveling? _____ ft/sec. b. What is the starting height of the falcon? _____ ft	Solve using the quadratic formula. Throwing a ball out of a window in a downward direction can be modeled by $h(t) = -16t^2 - 20t + 16$, where h is the height of the ball off the ground and t is time in seconds. How long does it take for the ball to hit the ground?

Quadratic Functions

Lesson 3: Solving Using the Quadratic Formula

Think Dots

Directions: In your groups, each student will take turns rolling the die. Answer the corresponding question on the record sheet. You may discuss your answers with your group members before recording your answer. You must answer all of the questions.

⚀	⚁	⚂
Define vertex, minimum value, maximum value, and axis of symmetry. Sketch two parabolas, one that opens upward and one that opens downward, and label these features.	Solve the following quadratic function using the quadratic formula. Check your solutions. $$h(t) = 4t^2 - 3t + 12$$	Write a program in your graphing calculator to use the quadratic formula to solve a quadratic equation.
⚃	⚄	⚅
A falcon diving from the top of a 200-foot building at a velocity of 100 feet/sec to catch its prey can be modeled with the function $h(t) = -16t^2 - vt + s$, where $h(t)$ represents the height in feet at t seconds. Plug in the two variables and use the quadratic formula to calculate how long it would take the falcon to reach its prey on the ground.	A cliff diver performing a dive can be modeled using a quadratic function. Complete a table of data to model this situation using estimated heights and times for the diver standing at the top of the cliff, jumping, and then hitting the water. Then use quadratic regression to write the equation that represents your data. Make sure your data make sense.	Write a description of a quadratic model that could be represented by the following equation. Pose a question that could be answered using the quadratic equation. $$y = -16x^2 - 60x + 58$$

Quadratic Functions

Lesson 3: Solving Using the Quadratic Formula

Think Dots Record Sheet

Circle worksheet chosen: ★

Quadratic Functions

Lesson 3: Solving Using the Quadratic Formula

Exit Slip

1. Write the quadratic formula.

2. Explain why the quadratic formula can be used to solve any quadratic function.

3. Explain why there may be two solutions to a real-life quadratic function application, but one answer can sometimes be eliminated.

Lesson 4: Solving by Factoring
Checkerboard Small-Group Activity Lesson Plan

Purpose: Students will work in small groups (in pairs or groups of three) to put together the checkerboard puzzle that reviews identifying solutions to linear equations and identifying the slope and y-intercept from an equation.

Prerequisite Knowledge	Materials Needed
Students should know: • how to find factors of numbers, • how to use FOIL to multiply binomials, • the exponent rules for multiplication, • how to combine like terms, and • the integer multiplication rules.	• One copy of the checkerboard for each small group of students, cut into the squares • Scrap paper for solving equations

Lesson
➤ Teachers may want to use random grouping. Each group should receive a packet of the puzzle pieces that are in no particular order.

➤ Students are asked to put together the checkerboard by matching pieces that have questions and answers that go together. There is no particular message revealed when the puzzle is completed.

➤ Students may notice that there are puzzle pieces that are blank on one side of the piece. They may figure out that these pieces belong on the left or right border of the puzzle.

Discussion Questions
➤ What does FOIL stand for? How do you multiply binomials?

➤ What is the rule for multiplying exponents?

➤ What are the rules for multiplying integers?

➤ What happens when you multiply $(x+1)(x-1)$? Why is the answer another binomial instead of a trinomial?

$2x^2+x-10$ x^2-2x-3 $(x+1)(x-3)$ **D** x^2+3x+2 $(x+1)(x+2)$	x^2-4x-3 $(x-3)(x-2)$ x^2-5x+6 **X** x^2-1 $(x+1)(x-1)$	$12x^2-6$ $4x^2-1$ **W** $(x-1)(x-8)$	$4x^2-2x-4$ $(2x-1)(2x+1)$ $4x^2-1$ **S** $2x^2-x-1$ $(2x+1)(x-1)$
$(x+1)(x+2)$ $3x^2-7x-20$ $(3x+5)(x-4)$ **R** $2x^2-8x-10$ $(x-5)(2x+2)$	$(x+1)(x-1)$ $x^2-8x-20$ $(x+2)(x-10)$ **L** $4x^2-6x-4$ $(2x-4)(2x+1)$	x^2-9x+8 $4x^2-4$ **M** $4x^2-2x-30$ $(2x+5)(2x-6)$	$(2x+1)(x-1)$ $(4x-4)(x+1)$ $4x^2-4$ **E** x^2-4 $(x+2)(x-2)$
$(x-5)(2x+2)$ x^2+8x+7 $(x+7)(x+1)$ **K** $15x^2-7x-2$ $(5x+1)(3x-2)$	$(2x-4)(2x+1)$ x^2-x-30 $(2x-4)(3x+5)$ **N** x^2-x-30 $(x-6)(x+5)$	$6x^2-2x-20$ $(2x-4)(3x+5)$ $(2x+5)(2x-6)$ **A** $4x^2-8x+4$ $(2x-2)(2x-2)$	$2x^2-12x+16$ $(2x-4)(x-4)$ $(x+2)(x-2)$ **H** $x^2+2x-15$ $(x+5)(x-3)$
$(5x+1)(3x-2)$ $2x^2+6x-56$ $(2x+7)(2x-8)$ **B** $x^2-9x+18$ $(x-3)(x-6)$	$(x-6)(x+5)$ **C** $5x^2-9x-2$ $(x-2)(5x+1)$	$(2x-2)(2x-2)$ $9x^2-1$ $(3x+1)(3x-1)$ **V** $10x^2+23x-5$ $(2x+5)(5x-1)$	$(x+5)(x-3)$ $x^2-4x-45$ $(x+5)(x-9)$ **U** $4x^2+2x-2$ $(2x-1)(2x+2)$
$(x-3)(x-6)$ $4x^2-20x+25$ $(2x-5)(2x-5)$ **D** $x^2-6x+18$	$(x-2)(5x+1)$ **I** $5x^2-20$	$(2x+5)(5x-1)$ $9x^2+6x+1$ $(3x+1)(3x+1)$ **Z** $4x^2-4x+4$	$(2x-1)(2x+2)$ $6x^2-2x-20$ $(3x+5)(2x-4)$ **T** $2x^2-8x-10$

Lesson 4: Solving by Factoring
Agenda

$$x^2 + bx + c$$

Imperatives (You must do all three of these.)

1. Complete the following sentence: When factoring a trinomial in the form $x^2 + bx + c$, we first need to find factors of _____ that add to _____ (use letters from the general equation).

2. Match the trinomial with the correct factorization. Draw a line to connect each match.

   $x^2 + 3x + 2$ $(x-1)(x+2)$

   $x^2 - 3x + 2$ $(x-1)(x-2)$

   $x^2 - x - 2$ $(x+1)(x+2)$

   $x^2 + x - 2$ $(x+1)(x-2)$

3. Factor each trinomial.

   a. $x^2 + 7x + 12$ e. $x^2 - 3x - 18$

   b. $t^2 + 8t + 12$ f. $n^2 + 5n - 24$

   c. $m^2 - 3m - 10$ g. $x^2 - 33x - 280$

   d. $b^2 - b - 20$

Zero-Product Property
Let a and b be real numbers.
If $ab = 0$, then $a = 0$ or $b = 0$.

Negotiables (You must do at least one of these.)

1. Factor the equations, if necessary, and then apply the zero-product property to find the zeros/solutions. Check your answer using a graphing calculator.

    a. $(x+12)(x+7) = 0$

    b. $(b-2)(b+4) = 0$

    c. $x^2 - 17x + 30 = 0$

    d. $n^2 - 20n + 19 = 0$

    e. $z^2 + 65z + 1000 = 0$

    f. $x^2 + 16x = -15$

2. Consider a rectangle that has a side length of $(x-5)$ and an area defined by $A = x^2 - 15x + 50$.
    a. Use factoring to find the expression for the length of the other side.

    b. If the area of the rectangle is 150 units, what are the possible values of x? If you found two answers, explain why one of the answers can be eliminated.

    c. What are the dimensions of the rectangle?

Options (You may choose to do this problem.)

1. When $x^2 + 3x - 40$ is factored, why can $(x+5)(x+8)$ and $(x-5)(x-8)$ be eliminated when testing possible factorizations?

Quadratic Functions

Lesson 4: Solving by Factoring

Agenda

$$ax^2 + bx + c$$

Imperatives (You must do all three of these.)

1. Complete the following sentence:

   When factoring a trinomial in the form $ax^2 + bx + c$, we first need to find factors of
   ____ • ____ that add to ____ (use letters from the general equations).

2. Match the trinomial with the correct factorization. Draw a line to connect each match.

   $2x^2 - 5x - 3$ $(3x-1)(2x+1)$

   $6x^2 + x - 1$ $(2+x)(-3-2x)$

   $-3x^2 + 2x + 1$ $(2x+1)(x-3)$

   $-2x^2 - 7x - 6$ $(-x+1)(3x+1)$

3. Factor each trinomial.

   a. $3x^2 - 17x - 6$ e. $8x^2 + 2x - 3$

   b. $3t^2 + 16t + 5$ f. $6n^2 - 11n - 2$

   c. $5m^2 - 9m - 2$ g. $2x^2 + 19x - 10$

   d. $4b^2 - 26b + 42$

Quadratic Functions

Negotiables (You must do at least one of these.)

1. Factor the equations, if necessary, and then apply the zero-product property to find the zeros/solutions. Check your answer using a graphing calculator.

   a. $(2x+3)(x-7)=0$

   b. $(4b-5)(2b-1)=0$

   c. $3n^2-37n+44=0$

   d. $14z^2-15z+4=0$

   e. $3n^2+34n+11=0$

   f. $6z^2+29z-5=0$

2. The width of a piece of carpeting for a living room is 2 feet less than twice its length. The area of the carpet is 180 square feet. What are the dimensions of the carpet?

Options (You may choose to do this problem.)

1. When Katie kicks a soccer ball, it can be modeled by the function $h(x)=-16x^2+40x$, where h is the height of the ball in feet and x is the time in seconds after it is kicked.

   a. Find $h(0)$ _____. What does this value mean?

   b. How many times is the ball at a height of zero?

   c. What is the highest the ball reaches?

   d. How long did it take to reach this height?

Lesson 4: Solving by Factoring
Tic-Tac-Toe Board

Directions: Choose three problems in a row, in a column, or on a diagonal. Solve the problems and show your work on a separate sheet of paper. Circle the problems you choose on this page and then reference them on your work paper with the number from the square.

Quadratic Functions

1	**2**	**3**

Determine if the trinomials can be factored into binomials with integers.
 	
a. $x^2 + 2x - 4$
b. $2x^2 + 5x - 3$
c. $3x^2 - 8x - 2$
d. $6x^2 + x - 1$
	When an object is dropped or thrown, it can be modeled using vertical motion models. For the following two models, define the variables.
	
	a. Object dropped:
	    $h(t) = -16t^2 + s$
	b. Object thrown:
	    $h(t) = -16t^2 + vt + s$
		The height of a baseball thrown into the air can be modeled by the function $h(t) = -16t^2 + 70t + 6$, where $h(t)$ is the height of the ball in feet after t seconds. How long does it take for the baseball to reach its highest height, and what is that height?

4	**5**	**6**
Factor the trinomials, if possible: a. $3t^2 + 7t - 6$ b. $3x^2 + 2x - 8$ c. $4n^2 - 26n - 42$ d. $4d^2 - 2d - 3$ e. $p^2 - 2p - 1$	The height of a diver above the water during a dive can be modeled by the function $h(t) = -16t^2 + 8t + 8$, where $h(t)$ is the height in feet and t is the time in seconds. a. How high is the diver at the start of the dive? b. How long does it take for the diver to hit the water?	Determine if the trinomials can be factored into binomials with integers. a. $6x^2 - 4x - 5$ b. $3d^2 - 4d - 7$ c. $2x^2 - 4x - 1$ d. $4w^2 - 14w - 30$
7	**8**	**9**
Create a quadratic function that has zeros of: a. $x = -\dfrac{1}{2}$ and $x = \dfrac{3}{5}$. b. $x = \dfrac{4}{3}$ and $x = \dfrac{2}{3}$.	Solve the quadratic equations using factoring and the zero-product property. a. $0 = 15x^2 - 11x + 2$ b. $0 = 12x^2 - 2x - 2$	You want to put a decorative stone walking path along two adjacent sides of a water garden. The water garden measures 4 ft by 10 ft. How wide should the path be if there is enough decorative stone to cover 51 square feet?

Lesson 4: Solving by Factoring

Exit Slip

1. What does it mean to solve a quadratic equation?

2. Why is the quadratic equation $6 = x^2 + 2x - 9$ not ready to be factored?

3. Explain how to factor the quadratic equation $0 = 2x^2 - 5x - 3$.

Differentiating Instruction in Algebra 1 © Prufrock Press Inc. • Permission is granted to photocopy or reproduce this page for single classroom use only.

Lesson 5: Solving by Completing the Square

Agenda

$$x^2 + bx = c$$

Imperatives (You must do all three of these.)

1. Use the square-root property to solve the equations.

   a. $x^2 = 9$ d. $x^2 = 10$

   b. $x^2 = 36$ e. $x^2 = 81$

   c. $x^2 = 18$ f. $x^2 = -4$

2. Match the trinomial with the correct squared binomial. Draw a line to connect each match.

   $x^2 + 4x + 4$ $(x-3)^2$

   $x^2 - 6x + 9$ $(x-1)^2$

   $x^2 - 2x + 1$ $(x+2)^2$

   $x^2 + 10x + 25$ $(x+5)^2$

3. Write the steps needed to solve a quadratic by completing the square.

   a. _____

   b. _____

   c. _____

   d. _____

   e. _____

   f. _____

4. Determine the value that needs to be added to or subtracted from both sides of the equation.

   a. $x^2 + 4x = 6$

   b. $x^2 + 10x = 16$

   c. $x^2 - 6x = 2$

   d. $x^2 - 8x = 0$

Quadratic Functions

Negotiables (You must do at least one of these.)

1. Answer the following questions.

    a. When a quadratic function is in the form of $0 = x^2 + bx + c$, why is completing the square easiest when b is an even number?

    b. If you have a quadratic with an a value of 5, what do you have to do first before you can use the completing the square strategy?

2. Solve $0 = x^2 + 4x - 12$ using completing the square and the quadratic formula. Check your answers in the original equation.

Options (You may choose to do this problem.)

1. Considering graphing, completing the square, factoring, and the quadratic formula, explain which method would be the best for each equation.

    a. $y = 2x^2 - 3x - 2$

    b. $y = 5x^2 + 13x - 2$

    c. $y = .25x^2 - 3x$

    d. $y = x^2 + 6x + 4$

Quadratic Functions

Lesson 5: Solving by Completing the Square
Think Dots

Directions: In your groups, each student will take turns rolling the die. Answer the corresponding question on the record sheet. You may discuss your answers with your group members before recording your answer. You must answer all of the questions.

Quadratic Functions

Multiply the binomials.

a. $(x+1)(x-1)$

b. $(x-3)(x-3)$

c. $(x+3)(x+3)$

d. $(x-5)^2$

Factor the trinomials. Write your answer as a squared binomial.

a. $x^2 + 8x + 16$

b. $x^2 - 6x + 9$

c. $x^2 + 10x + 25$

What numbers for x make the equations true?

a. $x^2 = 25$

b. $x^2 = 49$

c. $x^2 = 100$

d. $x^2 = 15$

Complete each square by determining what value needs to be added to both sides of the equation. Put the number in both spaces.

a. $x^2 + 2x + \_\_\_ = 5 + \_\_\_$

b. $x^2 - 6x + \_\_\_ = 4 + \_\_\_$

c. $x^2 + 12x + \_\_\_ = 7 + \_\_\_$

d. $x^2 + 4x + \_\_\_ = 3 + \_\_\_$

Solve by completing the square. Check your answer using the original equation.

a. $x^2 + 8x = 0$

b. $x^2 - 4x = 5$

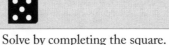

The price of a high-tech stock can be modeled with $P(t) = t^2 - 2t + 50$, where $P(t)$ is the price t weeks after the purchase.

a. What is the price at $t = 0$ and what does this answer represent?

b. What is the price at $t = 6$ and what does this answer represent?

c. What is the vertex and what does this point represent?

Lesson 5: Solving by Completing the Square
Think Dots

Directions: In your groups, each student will take turns rolling the die. Answer the corresponding question on the record sheet. You may discuss your answers with your group members before recording your answer. You must answer all of the questions.

Factor the trinomials. Write your answer as a squared binomial. a. $x^2 - 10x + 25$ b. $x^2 - 18x + 81$ c. $4x^2 - 12x + 9$	Solve each equation using the square-root property. Check your answers in the original equations. a. $(x+1)^2 = 25$ b. $(x-1)^2 = 49$ c. $(x+3)^2 - 10 = 15$	What numbers for x make the equations true? Check your answers. a. $x^2 - 3 = 22$ b. $x^2 + 2 = 51$ c. $x^2 - 6 = 30$
Complete each square by determining what value needs to be added to both sides of the equation. Put the number in both spaces. a. $x^2 - 6x + \underline{} = 2 + \underline{}$ b. $x^2 + 2x + \underline{} = 5 + \underline{}$ c. $x^2 + 10x + \underline{} = 1 + \underline{}$ d. $x^2 - 8x + \underline{} = 3 + \underline{}$	Solve by completing the square. Check your answer using the original equation. a. $x^2 + 8x + 6 = 0$ b. $2x^2 - 4x = 8$	A firework fired into the air can be modeled by $h(t) = -16t^2 + 80t$, where $h(t)$ is the height t seconds after launch. a. What is the maximum height reached by the firework? How long does it take to reach that height? b. Find the zeros. What do these two points represent?

Quadratic Functions

★

Lesson 5: Solving by Completing the Square
Think Dots

Directions: In your groups, each student will take turns rolling the die. Answer the corresponding question on the record sheet. You may discuss your answers with your group members before recording your answer. You must answer all of the questions.

⚀	⚁	⚂
Give an example of an equation that could be solved using completing the square and an equation that could be solved using another method.	A rectangular rug is $x + 8$ feet wide and $3x$ feet long. What are the dimensions if the area of the rug is 540 ft²?	a. Explain why $3x^3 - 7x + 6 = 0$ has no solutions. b. Give another example of a quadratic that has no solutions.
⚃	⚄	⚅
What numbers for x make the equations true? Check your answers. a. $(x-1)^2 - 2 = 23$ b. $(x+2)^2 - 1 = 22$ c. $(x+5)^2 + 3 = 10$	Solve by completing the square. Check your answer using the original equation. a. $x^2 + 4x + 2 = 0$ b. $2x^2 - 8x = 8$	Write a description of a quadratic model that could be represented by the following data. $$y = -16x^2 + 60x$$ Solve by completing the square and describe what the solutions mean.

Lesson 5: Solving by Completing the Square
Think Dots Record Sheet

Circle worksheet chosen:

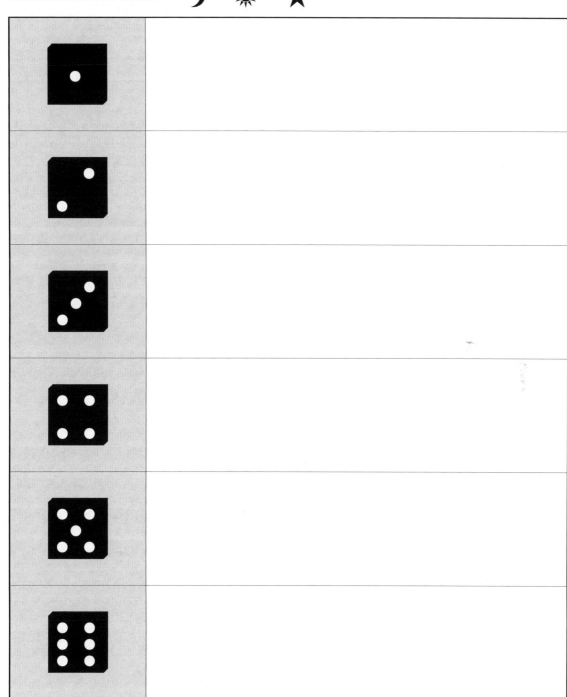

Quadratic Functions

Lesson 5: Solving by Completing the Square
Exit Slip

1. Given the equation $0 = 2x^2 - 4x - 8$, list the steps necessary to solve by completing the square. Solve the equation as you are listing the steps. Be sure to check your solutions.

a. _____

_____

b. _____

_____

c. _____

_____

d. _____

_____

e. _____

_____

f. _____

_____

Lesson 6: Wrap Up and Assessment

TriMind Activity

Learning Goal: Understand that there are several ways to solve quadratic functions that are found in everyday situations.

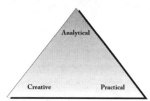

Directions: Select one of the following activities to complete.

Creative	Analytical	Practical
Write a children's book that discusses a quadratic function in real life. You need to include pictures and the solutions and what they mean. Your final product will be either a bound book or a large poster board with the pages in order.	Create a step-by-step procedure for solving quadratics using graphing, the quadratic formula, factoring, and completing the square. Your final product will be a typed listing of each step required for each method, including an explanation of each step.	Show how quadratic functions are used in everyday life. Create some examples for students to understand. Your final product can be a poster board, story, newspaper article, or research-type paper.

Quadratic Functions

Name: _____ Hour/Block: _____ Date: _____

Lesson 6: WRap Up and Assessment
TriMind Activity Rubric

Creative	Outstanding (5 points)	Good (3 points)	Poor (0 points)	Points Earned
Creativity	The final product was very creative.	The final product showed some creativity.	The final product lacked creativity.	
Real-life application	The application was realistic and original.	The application was original but not realistic.	The application was neither realistic nor original.	
Neatness	The product was very neat.	The product could have been neater but was readable.	The product was hard to read.	
			Total: _____ /15= _____ %	

Comments:

Analytical	Outstanding (5 points)	Good (3 points)	Poor (0 points)	Points Earned
Completeness	The final product had all of the required steps.	The final product had the majority of the required steps.	The final product was missing some steps.	
Understandable	The steps were understandable.	The steps were confusing.	The steps did not make sense or were in the wrong order.	
Neatness	The product was very neat.	The product could have been neater but was readable.	The product was hard to read.	
			Total: _____ /15= _____ %	

Comments:

Practical	Outstanding (5 points)	Good (3 points)	Poor (0 points)	Points Earned
Creativity	The final product was very creative.	The final product showed some creativity.	The final product lacked creativity.	
Realistic	The real-life applications were realistic.	Some applications were not realistic.	The applications were not realistic.	
Neatness	The product was very neat.	The product could have been neater but was readable.	The product was hard to read.	
			Total: _____ /15= _____ %	

Comments:

Lesson 6: Wrap Up and Assessment
RAFT Activity

Congratulations, our unit is nearing completion! You will now have the opportunity to demonstrate your understanding of one key element of the unit by choosing one of the following RAFT activities. If you would like to propose a change to any of the roles, audiences, or formats that are listed, you must have your idea approved before you start your RAFT. In addition, any student choosing the "student choice" option at the bottom must complete a Student Choice Proposal Form.

Role	Audience	Format	Topic
Vertex	Axis of symmetry	News story	Symmetry is pretty cool, but vertex is the king.
Quadratic function	Graphing calculator	Instruction manual	How do I set this window in my graphing calculator so I can see myself?
Catapult	NASA scientist	Billboard	I love launching stuff.
Factored form of an equation	Solutions	Children's book or storyboard	The book *Factoring for Dummies* was just published.
Solution	Other solutions	Advice column	What is this quadratic formula thing?
Quadratic equation	Linear equation	Bumper sticker	Quadratics rule!
Perfect square	Prime number	Morning announcements	What's a square got to do with anything? (completing the square)
Student choice	Student choice	Student choice	Student choice

Quadratic Functions

Lesson 6: Wrap Up and Assessment
RAFT Activity Rubric

Criteria	Excellent (2 points)	Acceptable (1 point)	Poor (0 points)	Points Earned
Role	The student assumed the appropriate role and wrote from the proper perspective.	The student assumed the role but the perspective was not correct.	The role did not appear related to the assignment.	
Audience	The audience was appropriately addressed.	The audience was addressed but not convincingly.	The intended audience was unclear.	
Format	The student clearly demonstrated an understanding of the format.	The student did not use the format well.	The format was not correct.	
Topic	The student demonstrated an understanding of the topic and communicated it well. The student used sample problems.	The student demonstrated some understanding of the topic. Sample problems were used sparingly.	The student clearly did not understand the topic. The student used very few sample problems.	
Creativity	The student demonstrated original thinking and entertained the audience.	The ideas presented were not original but the audience was entertained.	There was no creativity and the final product was not entertaining to the audience.	
Neatness	The final product was easy to read and understand.	Most of the product was neat and easy to read or understand.	The final product was not easy to read or understand.	
			Total: _____ /12= _____ %	

Comments:

Quadratic Functions

Quadratic Functions RAFT Student Choice Proposal Form	Name:
	Teacher's Approval:
	Date:

Proposals for Student Choices

Role:

Audience:

Format:

Topic:

- -Cut Here- -

Quadratic Functions RAFT Student Choice Proposal Form	Name:
	Teacher's Approval:
	Date:

Proposals for Student Choices

Role:

Audience:

Format:

Topic:

Lesson 6: Wrap Up and Assessment
When Marshmallows Fly: Catapult Activity Lesson Plan

Introduction

Quadratic functions are found in many everyday situations. Building catapults and launching marshmallows to hit a target will be interesting and fun for students and will assess student understanding of collecting data and developing a quadratic equation to model those data. Student teams will then use that equation to determine where to place the catapult in order to hit a predetermined target.

Common Core State Standards Addressed:	Big Ideas
• A.CED.2 • F.IF.4 • F.IF.7a • F.BF.1	• Quadratics can model many real-world situations.
	Essential Questions • What information do I need to write an equation to model a quadratic function?

Unit Objectives

As a result of this unit, students will know:
➤ how to work cooperatively in a team,
➤ how to gather data and use them to develop a quadratic equation,
➤ how to use data to make predictions,
➤ how to graph data,
➤ how to evaluate performance and make adjustments, and
➤ how to present results to an audience of peers.

Products/Skills Assessed

➤ Data collection
➤ Equation construction
➤ Graph of data/equation
➤ Ability to hit target
➤ Oral presentation

Launch

In cooperation with a history teacher, talk about how catapults have been used throughout history in various battle situations. Science teachers may have information on building catapults with ordinary household items. There are also multiple Internet sites with instructions.

Project Map—Key Steps

> ➤ Catapult research is completed
> ➤ Catapult design is completed and approved
> ➤ Catapults are constructed and ready for trial runs
> ➤ Catapults are adjusted based on trials
> ➤ Data are collected and equation is developed
> ➤ Graph of equation is completed
> ➤ Launch day is held
> ➤ Presentation of experiences is given

Quadratic Functions

Lesson 6: Wrap Up and Assessment
When Marshmallows Fly: Catapult Activity

Description: In this project, your group will use a catapult to launch marshmallows provided by your teacher. You will build a catapult, collect data, and write a quadratic function that represents your catapult's results. You will then use the equation to determine where to place your catapult in order to hit a predetermined target. A portion of your final grade will be based on your accuracy.

Based on the above description, what do you expect to learn from this project?

Procedure

➤ Research catapult design on the Internet. You must make your catapult using common household items.
➤ Make a blueprint of your catapult and list the supplies you will need.
➤ Get approval from your teacher for your design.
➤ Determine the team member roles needed to successfully complete this assignment. Assign roles to group members:

Role/Responsibility: _____

Student Assuming the Role:_____

Role/Responsibility: _____

Student Assuming the Role:_____

Role/Responsibility: _____

Student Assuming the Role:_____

Role/Responsibility: _____

Student Assuming the Role:_____

➤ Build your catapult with the supplies listed on your blueprint.
➤ Test your catapult.
➤ Develop a quadratic equation that models your catapult's launch data.
➤ Launch marshmallows to hit the target.

Quadratic Functions

Lesson 6: Wrap Up and Assessment

When Marshmallows Fly: Catapult Activity Data Collection

Your group must collect data in order to write a quadratic equation that models your catapult's launching capabilities. You must determine what data need to be collected and what you will do with those data.

1. Write the three forms of quadratic equations and describe what information is readily available from each form.
    a. Standard form:

    b. Vertex form:

    c. Factored form:

2. What information will be needed in order to write an equation to model your catapult?

3. What information will you gather in your catapult trial runs?

4. Describe how you will gather the data.

5. Summarize the data you collected and include a graph.

6. Write an equation that models your catapult.

Lesson 6: Wrap Up and Assessment

When Marshmallows Fly: Catapult Activity Project Summary

1. What was the best part of this project? What did you like best?

2. What was your least favorite part of this project?

3. How well did your team members work together? Describe the roles taken on by team members and how the team member performed.

4. Describe something that you learned while completing this project.

5. If you did this project again, what would you do differently?

6. How well did your team do in hitting the target? What contributed to your success or failure?

7. What features of your catapult could be changed if you wanted to change the flight pattern of the marshmallow?

Quadratic Functions

Lesson 6: Wrap Up and Assessment
When Marshmallows Fly: Catapult Activity

Criteria	Excellent (5 points)	Acceptable (3 points)	Poor (0 points)	Points Earned
Blueprint of catapult	Blueprint was accurate and detailed.	Blueprint included most elements but was not detailed.	Blueprint was incomplete.	
Data collection	Well-developed data-collection techniques were used.	Data-collection techniques resulted in data that were mostly reliable.	Data-collection techniques resulted in data that were not reliable.	
Equation	The equation was accurate based on the data.	The equation looked reasonable but did not include all data.	The equation did not fit the data.	
Graph	The graph included a title, scale, and labels and was accurate.	The graph was missing one element but was accurate.	The graph was missing two or more elements and was not accurate.	
Hit the target	The marshmallows hit the target.	The marshmallows landed within the target area.	The marshmallows landed outside of the target area.	
Group participation	All group members participated equally.	Group members participated, but there was clearly one person making the decisions.	The group could not work together.	
Neatness of project summary	Summary was easy to read and was complete.	Summary was incomplete or was not very neat.	Summary was incomplete and was not very neat.	

Total: _____/28= _____%

Comments:

Lesson 6: Wrap Up and Assessment
Exit Slip

1. Discuss three significant things that you learned about quadratic functions in this unit.

2. In what areas are you still struggling related to solving quadratic functions?

Lesson 7: Who Uses Mathematics?
Structured Academic Controversy Lesson Plan

Introduction

Do your students like to argue? Using a structured academic controversy can enable students to pick a side of an issue, research it, present their side, and then have constructive discourse with their peers. Constructive controversy is an instructional strategy that organizes students into small groups to develop and argue a position on an issue (Johnson et al., 2000). The issue included in this lesson plan asks students to argue whether mathematics and algebra are used in everyday life and future careers. Any relevant issue related to mathematics could be used instead.

Launch

Questions to launch this activity include:
➤ What careers require mathematics skills?
➤ Will you use your math skills after high school?
➤ What math do you use on a daily basis?
➤ Should students be required to take mathematics courses in high school?

Lesson

There are varying ways that a structured controversy can be planned, depending on the students in your classroom. One suggestion is to put students into groups of four. Each student receives a copy of the student worksheet. The groups of four are then split into pairs, and each pair takes one side of the issue. Each pair researches its side of the issue and records the three strongest arguments that support that side. The pairs then present their arguments. The students may not disagree with anything said; they may only clarify at this point. The pairs switch roles and repeat the process. After both pairs have presented, they can then respectfully challenge the other pair's position.

If time permits, the pairs can switch sides of the issue and briefly discuss the strengths of the other pair's presentation. Students can then synthesize a position on which they can all agree. They can write a consensus statement that will summarize what they have discussed and concluded as a group.

Summary

There are many ways to conclude this lesson. One way is to designate each side of the classroom as one side of the issue. Students then go to the side that represents the side of the issue they support. Students then share their strongest arguments for the view they support.

Example

The following is an example of what this might look like.
➤ **The issue to be discussed:** Use of mathematics and algebra in everyday life and future careers
➤ **One position:** I will never use the mathematics and algebra concepts I learned in high school in everyday life and a future career.

> **Opposite position:** I will use the mathematics and algebra concepts I learned in high school in everyday life and a future career.

As a result of this activity, students will be able to:
> argue a point in front of their peers,
> support their side of an argument with facts and reasonable thoughts,
> describe careers that require the use of algebra and mathematics,
> understand other students' positions,
> disagree with others' positions politely and with respect,
> accept criticism of their ideas,
> change their minds about their original position on the issue based on the evidence presented by the other side, and
> write a consensus statement that clearly articulates their position on an issue.

Lesson 7: Who Uses Mathematics?
Structured Academic Controversy Rubric

	Outstanding (2 points)	Good (1 point)	Poor (0 points)	Points Earned
Development of arguments	The arguments were well developed and articulated.	The arguments lacked substance.	The arguments were weak.	
Listening skills	The student listened to others' positions and recorded best opposing views.	The student did not consider others' positions and did not articulate the best arguments from opponents.	The student did not listen to others' positions and did not present the best opposing arguments.	
Consensus statement	The statement was well written.	The statement lacked substance.	The statement was very weak.	
			Total: _____/6= _____%	

Comments:

Quadratic Functions

Lesson 7: Who Uses Mathematics?

Structured Academic Controversy

The side you are defending: _____

1. Our three best arguments to support our side:

2. Notes and important points after hearing the opposing view:

3. The best three arguments for the opposing view:

4. The consensus statements related to our issue:

Quadratic Functions

Answer Key

Unit 1: Introduction to Functions and Relationships

Introduction to Functions and Relationships Preassessment

1. a. Domain: $\{1,2,4,7\}$, Range: $\{3,4,5\}$; b. Domain: $\{-2,-1,0,1,2\}$, Range: $\{-3,1,3,4,5\}$
2. a. is a function; b. is not a function; c. is a function; d. is a function
3. A function will pass the vertical line test.
4. a. $f(-1)=5$; b. $f(5)=-3$; c. $g(2)=-1$; d. $x=\dfrac{5}{4}$
5. a. $f(x)+g(x)=-2x+3$; b. $g(x)-f(x)=-6x+5$; c. $f(x)-g(x)=6x-5$
6. independent variable = hours worked; dependent variable = total cost; $f(t)=50+45t$

Lesson 1: Introduction: Agenda

Imperatives

1. a. is a function, Domain: $\{1,2,3,4\}$, Range: $\{2,3,5,7\}$; b. is a function, Domain: $\{1,4,6,8\}$, Range: $\{3,5,8\}$; c. is not a function
2. a. is not a function; b. is a function, Domain: $\{-3,-2,-1,0,1,2,3,4,5\}$, Range: $\{-5,-3,-2,1\}$; c. is a function, Domain: $\{-3,-2,-1,0,1,2,3,4,5\}$, Range: $\{1\}$
3. Graphs that are functions will pass the vertical line test.

Negotiables

1. a. A person and his or her birth date is a function because each person has just one birth date; b. First name and last name is not a function because the first name Katie could be matched with many different last names; c. City and zip code is not a function because a city could have more than one zip code; d. Last name and first name is not a function because the last name Smith could be matched with many different first names.
2. Answers will vary.
3. Given a graph, draw vertical lines, and if any of the lines touches the graph more than one time, the graph does not represent a function. The line should be able to be drawn in any place on the function. Graphs will vary.

Options

1. A relation or a function can be represented by a table, a graph, coordinate pairs, mapping diagrams, or an equation.
2. Answers will vary.

Lesson 1: Introduction: Exit Slip

1. A function is a special type of relation in which each domain item is paired with exactly one range item.
2. If a vertical line can be drawn on a function in any location and the line only touches the function one time, the relation is a function.
3. Answers will vary. Possible answers: first table (2, 0); second table (-1, 2)

Lesson 2: Writing Functions: Agenda

Imperatives

1. a. slope = 2; b. slope = -6; c. slope = $\dfrac{1}{2}$; d. slope = 6

2. a. *y*-intercept = 3; b. *y*-intercept = 1
3. a. $y = 2x - 1$ is a function; b. $y = -3x - 2$ is a function
4. a. Cost depends on weight, independent variable = weight, dependent variable = cost; b. Cost depends on days rented, independent variable = days rented, dependent variable = cost; c. Cost depends on pounds, independent variable = pounds, dependent variable = cost

Negotiables
1. a. independent variable = hours, dependent variable = cost; b. $f(t) = 150t$
2. a. independent variable = number of rides, dependent variable = cost; b. $f(r) = 5 + 2r$
3. a. $y = 3x - 2$; b. $y = 0.5x + 2$
4. a. independent variable = weeks, dependent variable = money in her bank; b. $y = 200 + 20w$; c. \$260; d. \$1,240; e. \$5,400

Options
1. Answers will vary. Possible answers include: a. A child puts \$2.50 in a piggy bank each week; b. A bank account starts with \$50 and \$2 is added each day; c. There are 25 coins in the jar and two more are added each day; d. To save for an upcoming vacation, Erin puts \$100 in an account and adds \$5 per week.

Lesson 2: Writing Functions: Tic-Tac-Toe Board
1. $f(x) = 5x - 3$
2. a. $f(g) = 3.59g$; b. $f(w) = 1,250w$; c. $f(x) = 1.89x$
3. a. independent variable = hours worked, dependent variable = paycheck total; b. independent variable = hours studying, dependent variable = grades received
4. $f(t) = 25 + .30t$, independent variable = texts, dependent variable = total bill
5. Answers will vary; Katie earns a flat rate of \$75 plus \$3 for each customer she signs up for new phone service; independent variable = number of customers; dependent variable = total pay
6. a. $f(p) = 1.69p$; b. $f(p) = .10p$; c. $f(m) = 125 + 57m$
7. a. independent variable = pounds, dependent variable = total cost, $f(p) = .59p$; b. independent variable = weeks, dependent variable = total cost, $f(w) = 20w$; c. independent variable = hours, dependent variable = total cost, $f(h) = 25h$
8. a. independent variable = inches, dependent variable = cost; b. independent variable = pounds; dependent variable = cost
9. missing values: *x*-value = 1, *y*-value = -53

Lesson 2: Writing Functions: Exit Slip
1. Answers will vary. The dependent variable depends on the independent variable. The independent variable is graphed on the *x*-axis, and the dependent variable is graphed on the *y*-axis. Independent variables can stand alone while dependent variables depend on other factors.
2. Answers will vary. Erin receives a weekly allowance of \$35; $f(w) = 35w$

Lesson 3: Evaluating Functions and Function Notation: Agenda

Imperatives
1. a. $f(1) = -1$; b. $f(0) = -5$; c. $g(-2) = -12$; d. $h(12) = 3$

2. a. $f(0) = 7$; b. $f(-2) = 9$; c. $f(-4) = 6$; d. $f(5) = 2$; e. $f(-7) = 2$ and $f(5) = 2$;
    f. $f(-7.5) = 1$

3. a. $h(3) = 28$, after 3 seconds the balloon is 28 feet off the ground; b. $h(10) = 84$, after 10 seconds the balloon is 84 feet off the ground; c. 9 seconds

Negotiables

1. a. $f(w) = 35 + .25w$; b. $f(c) = 1{,}500 + .10c$
2. a. $-x + 6$; b. $7x - 8$; c. $6x - 2$; d. $-6x + 19$
3. a. no, $f(0) = 7$; b. yes; c. no, $f(-1) = 11$; d. yes

Options

1. a. $f(x) = 5x - 2$; b. $g(x) = 3x^2 + 4$

Lesson 3: Evaluating Functions and Function Notation: Hexagon Puzzles

Moon	Sun	Star
A→domain: $\{-7, -2, 2, 5\}$	A→domain is input; range is output	A→Yes. Each state has just one governor.
B→range: $\{3, 5, 6, 7\}$	B→Draw a vertical line, and if the line touches the function in more than one place, it is not a function. The line should be able to be drawn in any location on the function.	B→Yes. Each person has just one birth date.
C→ <table><tr><td>x</td><td>−2</td><td>−3</td><td>−5</td><td>−8</td></tr><tr><td>y</td><td>3</td><td>6</td><td>7</td><td>9</td></tr></table>	C→Graph, table, equation, mapping diagram, and coordinate pairs	C→No. Many large cities have several zip codes.
D→ (mapping diagram: 1, 4, 6 → 2, 3, 5, 7)	D→domain is the inputs or x-values; range is the outputs or y-values	D→ No. People within the same family have the same last name.
E→No, cities and zip codes do not represent a function because some cities have more than one zip code.	E→ Answers to this may vary. The majority of cities are in one specific area code for telephone numbers. However, some large cities contain multiple area codes (e.g., Houston).	E→independent = time; dependent = fees
F→ Yes, states and their governors represent a function because each state has just one governor.	F→No. Each state has two U.S. Senators.	F→independent = pounds; dependent =cost
G →Yes, each person has just one social security number.	G→No. Someone with the first name of Robert could have various last names such as Smith, Jones, Sanford, and so forth.	G→independent = hours; dependent = pay

Unit 1: Introduction to Functions and Relationships

Moon	Sun	Star
H→Yes, each person has just one birth date.	H→ Yes. Each person has just one birth date.	H→independent = tickets; dependent = cost
I→slope = $\frac{1}{3}$	I→ $-\frac{1}{3}$	I→-2
J→slope = $\frac{2}{3}$	J→ $\frac{7}{4}$	J→-17
K→slope = 3; y-intercept = 6	K→slope = 3; y-intercept = -2	K→10
L→slope = -1; y-intercept = -5	L→slope= -3; y-intercept = 12	L→-25
M→independent = pounds; dependent = cost	M→Answers will vary. Example: allowance and the amount in a bank account after any number of weeks. Independent = number of weeks; dependent = total amount	M→ Answers will vary.
N→independent = hours; dependent = total pay	N→independent = height of the tree; dependent = cost	N→Answers will vary. Possible answer: $(1,2),(1,5)(1,6)$
O→independent = time spent babysitting; dependent = fees	O→independent = number of children; dependent = babysitting fees	O→Answers will vary. Possible answer: $(1,2),(2,3)(4,5)$
P→independent = number of tickets; dependent = cost	P→independent = number of hours; dependent = cost	P→ The coordinate pair combinations could include $(2,2),(-2,2),(3,3),(-3,3)$. It is a function, and it passes the vertical line test.
Q→-17	Q→13	Q→10
R→10	R→14	R→20
S→-2	S→-2	S→-3
T→-25	T→39	T→36
U→ $2x+6$	U→ $-4x+7$	U→25
V→ $-x-2$	V→ $x+2$	V→0
W→ $-x+9$	W→ $x+9$	W→28
X→ $-3x-12$	X→ $3x-12$	X→-6

Lesson 3: Evaluating Functions and Function Notation: Exit Slip

1. a. $f(0)$ represents the starting value of $50 at time zero; b. 250; c. $f(8) = 250$
2. The g-function must be subtracted from the f-function. Like terms must be combined. $f(x) - g(x) = -5x + 3$

Lesson 4: Wrap Up and Assessment: Think Dots

Moon

1.

x	−4	−2	2	4
y	7	5	5	7

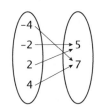   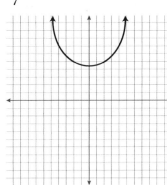

2. The missing x-value could be any number that hasn't already been used. The missing y-value could be any number. Example: 3, 12

3. Students should draw a graph that is a function and draw a vertical line and then draw a graph that is not a function and draw a vertical line and show two intersection points.

4. a. independent variable = number of pounds, dependent variable = total price of the candy; $f(x) = 1.89x$

5. a. $f(0) = 2$; b. $f(-2) = -2$; c. $f(-8) = 14$; d. The $f(x)$ function can be put into the graphing calculator, and one would then find the answers in the table.

6. a. $f(x) + g(x) = 6x + 4$; b. $f(x) - g(x) = -2x - 6$; c. $2f(x) + 4g(x) = 20x + 18$

Sun

1.

x	−4	−2	2	4
y	7	5	5	7

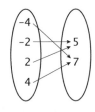   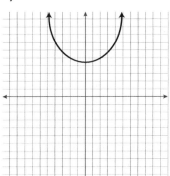

2. Answers will vary. An example of a function is: $(0,10),(1,8),(2,6),(3,5),(4,3)$; an example of a nonfunction is: $(0,15),(1,3),(2,12),(3,10),(4,5),(4,6)$

3. If a vertical line intersects the relation's graph in more than one place, the relation is not a function. Answers will vary on the graphs.

4. a. independent variable = hours, dependent variable = amount of rain; b. $f(x) = 0.75x$

5. a. $f(0) = -4$; b. $f(-2) = 0$; c. $f(5) = -14$; d. answers will vary

6. a. $f(x) + g(x) = x + 4$; b. $f(x) - g(x) = 5x - 10$; c. $2f(x) + 4g(x) = -2x + 22$

Star

1. Answers will vary for the domain. D: $\{0 \le x \le 50\}$ and R: $\{0 \le y \le 350\}$

2. Answers will vary. An example of a function is: $(0,10),(1,8),(2,6),(3,5),(4,3)$; an example of a nonfunction is: $(0,15),(1,3),(2,12),(3,10),(4,5),(4,6)$

3. If a vertical line intersects the relation's graph in more than one place, the relation is not a function. Answers will vary on the graphs.

4. Answers will vary. One possible answer: Firewood costs \$100 per cord. a. independent variable = number of cords of wood, dependent variable = total cost; b. $f(x) = 100x$

5. a. $f(0) = -4$; b. $f(-2) = 0$; c. $f(5) = -14$; d. answers will vary

6. a. $f(g(5)) = -12$; b. $g(f(-1)) = 19$; c. $2f(x) + 4g(x) = -2x + 22$

Unit 2: Systems of Linear Equations

Preassessment

1. Not a solution

2. $(-3,7)$

3. $(3,-7)$

4. $(1,-1)$

5. Equations: $a + c = 443$, $5.50a + 3.00c = 1,566.50$; solution: 95 adult tickets and 348 children's tickets

6. Equations: $x + y = 60$, $0.05x + 0.02y = 0.03(60)$; solution: 20 ml of 5% and 40 ml of 2%

Lesson 2: Solving Systems Using Graphing: Tic-Tac-Toe Board

1. Answers will vary. Both equations must be solved for y in order to use the graphing calculator. Both equations are loaded into $y =$ and then graph the lines. The point where the two lines intersect is the solution.

2. Answers will vary.

3. Answers will vary. Any two equations that have a y-intercept at $(0,0)$ will satisfy the requirement. One possible solution: $y = 3x$ and $y = -2x$

4. Solution: $(-2,0)$

5. Answers will vary. One possible solution: $y = 2x$ and $x = 1$

6. Answers will vary. A system that has no solutions includes lines that have the same slope ($y = 2x + 1$ and $y = 2x - 3$). A system that has an infinite number of solutions includes lines that are the same ($y = -3x + 1$ and $2y = -6x + 2$)

7. Answers will vary.

8. Answers will vary. One possible solution: $y = 2x - 4$ and $y = \frac{1}{3}x + 1$

9. Answers will vary. One possible solution: $y = 4x - 2$ and $y = 2$

Lesson 2: Solving Using Graphing: Graphing Systems

Moon	Sun	Star
1. $\left(\frac{1}{2},0\right)$	1. $\left(\frac{1}{2},0\right)$	1. $(0,-1)$
2. $\left(-\frac{1}{2},-4\frac{1}{2}\right)$	2. $(-3,-1)$	2. $(-3,-1)$
3. $(-3,-1)$	3. $\left(\frac{3}{4},-\frac{1}{2}\right)$	3. $\left(\frac{3}{4},-\frac{1}{2}\right)$
4. $(1,3)$	4. Infinite number of solutions	4. Infinite number of solutions
5. No solutions	5. No solutions	5. $(1.2,-2.6)$
6. $(4,5)$	6. $(4,5)$	6. $(-0.8,-0.1)$

Lesson 2: Solving Using Graphing: Exit Slip

1. Answers will vary. Graphing is an easy method to use when both equations are written in the form $y = mx + b$ so that the equations can be plugged into the graphing calculator. Once the equations are put into the calculator, find the intersection of the two lines to identify the solution.

2. Answers will vary. One possible answer: $y = 2x - 7$ and $y = -3x + 1$

Lesson 3: Solving Using Substitution: Agenda

Imperatives

1. Answers will vary. a. consistent: a system with at least one solution ($y = 5x - 1$ and $y = 3x + 1$); b. inconsistent: a system that has no solutions ($y = 5x - 1$ and $y = 5x - 2$); c. dependent: a system with an infinite number of solutions ($y = 2x + 2$ and $3y = 6x + 6$); d. independent: a system with one solution ($y = 7x - 1$ and $y = -8x + 5$)

2. a. $(-2,3)$; b. $\left(\frac{1}{2},3\right)$; c. $(8,6)$

3. y = total cost and x = months, $y = 49 + 29x$ and $y = 25 + 37x$; a. 3 months; b. $136; c. Satellite Systems will be the cheapest at $223

Negotiables

1. v = vans and b = buses; $v + b = 9$ and $53b + 8v = 342$; 6 buses and 3 vans were used for the field trip

2. w = wakeboards and s = surfboards; $w + s = 85$ and $49w + 129s = 6,965$; 35 skateboards and 50 wakeboards were sold

3. y = total students and x = weeks; $30 + 2x = y$ and $10 + 3x = y$; 20 weeks

Options

1. A = -5; B = -3; C = 3

2. A = -4; B = 5; C = 17; D = -1; E = 2

Lesson 3: Solving Using Substitution: Exit Slip

1. Answers will vary. One or both of the equations need to be solved for one variable in order to use substitution. Take the equation that is solved for a variable and plug it into the other equation for that variable.

2. Answers will vary. Possible answers: $y = 2x - 1$ and $2x + 4y = 12$; $4x - 2y = 4$ and $x = 12 - y$

Lesson 4: Solving Using Elimination: Tic-Tac-Toe Board

1. $(4, 0)$
2. x = number of games and y = total cost; $y = 4 + 2x$ and $y = 2 + 2.5x$; 4 games and \$12 spent
3. x = number of CDs and y = number of DVDs; $15x + 19y = 465$ and $x + y = 27$; 12 CDs and 15 DVDs
4. d = dimes and q = quarters; $d + q = 28$ and $.10d + .25q = 5.05$; 13 dimes and 15 quarters
5. $(2, 1)$
6. $x = -2$; $y = -3$; $z = 3$
7. $(0, 4)$
8. x = hours worked by Erin and y = hours worked by Katey; $x = 2y + 1$ and $3y + x = 6$; Erin worked 3 hours, and Katey worked 1 hour
9. x = first number and y = second number; $x + y = 66$ and $x = 2y - 24$; $x = 36$ and $y = 30$

Lesson 4: Solving Using Elimination: Exit Slip

1. Answers will vary. The variables and constants in the equations need to be in the same order such that the equations can be added together in order to eliminate one of the variables. One or both of the equations can be multiplied by a constant so that one of the variables can be eliminated.

2. Answers will vary. Possible answers: $2x + 4y = 1$ and $-2x + 3y = 10$; $3x + 4y = 6$ and $2x - 5y = 10$

Lesson 5: Mixture Problems: Agenda

Imperatives

1. a. 0.10; b. 0.01; c. 0.0006; d. 1.35; e. 10; f. 0.25; g. 0.006; h. 1; i. 2; j. 0.025
2. a. $x = .50(10)$, 5; b. $x = .25(40)$, 10; c. $x = .75(100)$, 75; d. $x = .01(1006)$, 10.06; e. $10 = x(100)$, 10; f. $4 = x(50)$, 8%; g. $15 = .30x$, 50; h. $90 = .75x$, 120
3. Equations: $x + y = 60$, $0.05x + 0.02y = 0.03(60)$; solution: 20 ml of 5% and 40 ml of 2%

Negotiables

1. x = percentage of M&Ms; $7(.30) + 2(.20) = 9x$; $x = 28\%$
2. x = concentration of a new mixture; $4(.25) + 7(.38) = 11x$; $x = 33\%$
3. x = mg of a metal; $.40x + 8(1.00) = .65(8 + x)$; $x = 11.2$

Options

1. x = 50% mixture (ounces) and y = 19% mixture (ounces); $24(.25) = .50x + .19y$ and $x + y = 24$; $y = 19.4$ and $x = 4.6$
2. x = ounces of 30% nickel and y = ounces of 50% nickel; $x + y = 16.6$ and $.30x + .50y = 16.6(.40)$; $x = 8.3$ and $y = 8.3$

Lesson 5: Mixture Problems: Exit Slip

1. Answers will vary. In mixture problems, generally one equation is for the total number of units that are being mixed and the other includes the percentages applied to the units.
2. Answers will vary. One possible solution: A 10-pound bag of premium cat food contains 70% chicken protein. It is formulated using a 50% chicken protein mixture combined with an 80% chicken mixture. How many pounds of each mixture are needed to make the 10-pound bag?

Lesson 6: Wrap Up and Assessment: Real-Life Applications

1. d = dimes and n = nickels; $n+d=25$ and $.05n+.10d=2.00$; 10 nickels and 15 dimes
2. x = first number and y = second number; $x+y=90$ and $x-y=34$; 28 and 62
3. a = adult tickets and c = children's tickets; $a+c=288$ and $5a+3.50c=1,161$; 102 adult tickets and 186 children's tickets
4. x = 5% solution and y = 2% solution; $x+y=60$ and $0.05x+0.02y=0.03(60)$; 20 ml of 5% and 40 ml of 2%
5. t = bags of tulips and d = bags of daffodils; $10t+8d=121$ and $5t+12d=116.50$; $7 for daffodils and $6.50 for tulips
6. y = total cost and x = months; $y=180+40x$ and $y=100+60x$; 4 months and the cost is $340
7. v = vans and b = buses; $45b+12v=330$ and $v+b=11$; 6 buses and 5 vans
8. c = chickens and g = goats; $c+g=52$ and $2c+4g=128$; 40 chickens and 12 goats

Lesson 6: Wrap Up and Assessment: Think Dots

Moon

1. Answers will vary. One solution: $y=2x+4$ and $y=-3x+5$; no solutions: $y=2x+1$ and $y=2x-2$; infinite number of solutions: $y=2x+1$ and $2y=4x+2$
2. a. elimination; b. graphing; c. substitution
3. Multiply the second equation by 3 to eliminate the y-variable; multiply the first equation by 3 and the second equation by 2 to eliminate the x-variable.
4. $(-0.5,4)$
5. Answers will vary. One possible answer: A friend opens an account with $50 and adds $2.50 per week. Another friend opens an account with $25 and adds $3.50 per week. When will they have the same amount in their accounts?
6. The second equation needs to be solved for y. Both equations can then be loaded into the graphing calculator. Find the intersection of the two lines.

Sun

1. Answers will vary. One solution: $y=2x+4$ and $y=-3x+5$; no solutions: $y=2x+1$ and $y=2x-2$; infinite number of solutions: $y=2x+1$ and $2y=4x+2$
2. The second equation needs to be solved for y so that it can be loaded into the graphing calculator. $5x=2y-1$ can be rewritten $\left(\frac{5}{2}\right)x+\frac{1}{2}=y$.
3. Multiply the first equation by 3 and the second equation by 2 to eliminate the x-variable; multiply the first equation by 7 and the second equation by 3 to eliminate the y-variable.
4. $(-2,-6)$

5. Answers will vary. One possible answer: You are selling pizza for $1.50 a slice and hot dogs for $2 each. If you sold a total of 125 pizza slices and hot dogs and collected $225, how many of each did you sell?

6. x = hot dogs and y = hamburgers; $x + y = 128$ and $1.00x + 1.50y = 151$; 82 hot dogs and 46 pizza slices

Star

1. Answers will vary. One possible answer: A 25% solution is mixed with a 57% solution to form a 35% solution. If 90 liters of the solution was created, how many liters of each solution were mixed?

2. $0.30x + 1.00(10) = 0.60(x + 10)$; $x = \dfrac{40}{3}$ or 13.3

3. The equations need to be rearranged so that the variables and constants are lined up. Multiply the first equation by 5 and the second equation by 2 to eliminate the x-variable; multiply either equation by -1 to eliminate the y-variable.

4. An independent system has just one solution (example: $y = 3x + 1$ and $y = 5x - 4$). A dependent system has an infinite number of solutions (example: $y = 2x + 2$ and $3y = 6x + 6$).

5. $\left(-\dfrac{12}{5}, \dfrac{19}{5}\right)$ or (−2.4, 3.8)

6. $x = 4; y = -5; z = -5$

Unit 3: Exponent Rules and Exponential Functions

Exponent Rules and Exponential Models: Preassessment

1. a. linear; b. neither (quadratic); c. exponential; d. exponential
2. a. exponential (doubling growth rate); b. not exponential
3. Answers will vary. Exponential decay will be a curve that decreases from left to right and exponential growth will increase from left to right.
4. a. $2^5 = 32$; b. $5^9 = 1{,}953{,}125$; c. a^5; d. $4^6 = 4{,}096$; e. d^{15}
5. a. $6^2 = 36$; b. a^3; c. $\dfrac{2^2}{4^2} = \dfrac{1}{4}$; d. $\dfrac{9x^2}{16y^2}$; e. $\dfrac{5^4}{t^4} = \dfrac{625}{t^4}$
6. a. 1; b. 16; c. $\dfrac{1}{27}$; d. $\dfrac{1}{4{,}096}$; e. $\dfrac{1}{32d^5}$
7. a. Starting value = 450; b. Growth factor = 8% or 0.08; c. Time period = 4; d. Final value = y
8. $y = 650(1 + 0.04)^3$; $731.16

Lesson 1: Introduction: Exit Slip

1. Answers will vary. Square roots and squaring numbers are opposites of each other. If you square 3, you get 9. If you take the square root of 9, you get the 3.
2. a. 10, 12, 14; b. 512, 2048, 8192, 32768; c. 3906.25, 976.5625; d. 0.05, 0.005, 0.0005, 0.00005

Lesson 2: Exponential Growth: *One Grain of Rice* by Demi and Doubling Pennies Tracking Sheet

1	2	4	8	16	31
32	64	128	256	512	1,023
1,024	2,048	4,096	8,192	16,348	32,731
32,768	65,536	131,072	262,144	524,288	812,609
1,048,576	2,097,152	4,194,304	8,388,608	16,777,216	33,315,465
33,554,432	67,108,864	134,217,728	268,435,456	536,870,912	1,043,303,857

Lesson 2: Exponential Growth: Agenda

Imperatives

1. a. 16, 32, 64 (start with 1 and multiply by 2); b. 81, 243, 729 (start with 1 and multiply by 3); c. 1000, 10000, 100000 (start with 1 and multiply by 10); d. 62.5, 156.25, 390.625 (start with 1 and multiply by 2.5); e. 11.71875, 14.6484375, 18.31054688 (start with 6 and multiply by 1.25)

2. a. 0.25; b. 0.12; c. 0.36; d. 1.0; e. 1.02; f. 2.0; g. 0.1; h. 0.08; i. 0.01; j. 0.001; k. 0.0025; l. 0.005

3. a. linear, not exponential; b. exponential, growth rate is 3(1 + 200%); c. exponential, the growth rate is 2(1 + 100%)

Negotiables

1. a. 12288, 3, 2 (doubling or 100% increase); b. 4374, 18, 3 (tripling or 200% increase)
2. $y = 210,000(1 + 0.05)^{10}$; $342,068$
3. $y = 10,000(1 + 0.04)^{x}$; between 10 and 11 years: After 10 years, the account has $14,802, and after 11 years, it has $15,395.

Options

1. Answers will vary. In the long term, 4^{x} grows more quickly than x^{4}. If you graph it or look at the table, you see how quickly the exponential function grows.
2. Using $x = 10$, $4^{10} = 1,048,576$ and $10^{4} = 10,000$.

Lesson 2: Exponential Growth: Tic-Tac-Toe Board

1. a. linear (slope = $\frac{4}{3}$); b. exponential (doubling or 100%); c. neither (quadratic); d. exponential (20%); e. exponential (150%)
2. a. $1480.24; b. $1967.15; c. $2593.74
3. a. growth (90%); b. growth (10%); c. decay; d. decay; e. neither growth nor decay
4. Answers will vary. Look at the changes in the y-values in the table. The multiplication factor is the growth rate.
5. Tripling means a 200% increase; 11,664 after 6 years
6. a. 5 years = $1,914.42; b. 9 years = $2,326.99; c. 15 years = $3,118.39
7. Approximately $35,463; approximately 12 years
8. The second function is growing faster because it has a 20% growth rate compared to 10%; when $x = 10$, the first function has a higher value.
9. (1,10) and (4,80)

Lesson 2: Exponential Growth Exit Slip

1. $y = A(1+r)^x$; y = final value, A = initial or starting value; r = rate of growth; x = time period
2. $y = 5,600(1+0.025)^x$; approximately 6 years

Lesson 3: Exponential Decay: Agenda

Imperatives

1. a. 3, 1.5, 0.75 (start with 24 and multiply by 0.50); b. 0.1, 0.01, 0.001 (start with 100 and multiply by 0.10); c. 3.125, 0.78125, 0.1953125 (start with 200 and multiply by 0.25); d. 1.125, 0.28125, 0.0703125 (start with 72 and multiply by 0.25); e. answers will vary
2. $(1-.75); (1-.88); (1-.64); (1-0); (1-.09); (1-.14); (1-.90); (1-.92); (1-.99); (1-.97);$ $(1-.35); (1-.01)$
3. a. exponential, decay rate of 50%; b. linear; c. exponential, decay rate of 90%

Negotiables

1. a. $409.60, $1000, decay rate of 20%; b. 411.94 or 412, 1500, decay rate of 35%
2. $y = 30,000(1-.12)^x$; approximately $20,444
3. $y = 35(1-.15)^x$; approximately 21.49 milligrams

Options

1. Answers will vary. For the first function, $y = 10(.75)^x$, the y-values are decreasing as the x-values increase. For the second function, $y = -10(.75)^x$, the y-values are increasing as the x-values increase.

Lesson 3: Exponential Decay: Tic-Tac-Toe Board

1. Answers will vary. Growth models can include situations such as money deposited in an account or an antique car appreciating in value. Decay models can include situations such as a car decreasing in value or a population of a town decreasing.
2. $y = 25,000(1-.12)^x$; between 7 and 8 years
3. a. growth; b. decay of 10%; c. growth; d. decay of 3%; e. decay of 99%
4. Answers will vary. As long as the x-values are increasing by one, you can take the y-values and divide them by the previous number in the table. Continue to do this for the entire table. If you get a constant multiplier, 1 minus the decay rate is the constant multiplier.
5. Answers will vary. The first equation is decaying at a 20% rate; the second equation is decreasing at a 25% rate. Therefore, the second equation is decaying faster. Only b values less than the number 1.0 will result in a decay function.
6. $y = 95(.75)^x$; on the 8th bounce, the ball will be under 10 cm
7. $125,000
8. Approximately 1,437; approximately 28 years
9. 4

Lesson 3: Exponential Decay: Think Dots

Moon

1. $(1-.25)$ or $(.75)$ and $(1-.50)$ or (0.50)
2. Answers will vary. Possible solutions: $y = 4(1-.10)^x$; $y = 250(1-.90)^x$; $y = (.25)^x$
3. $f(x)$ is the final value or answer; A is the starting value; in $1+r$, r represents the growth rate; x is time
4. a. $y = 250(1+.06)^x$; b. $y = 5(1-.02)^x$; c. $y = 300(1+1.0)^x$

5. a. growth; b. decay; c. decay; d. growth
6. a. 46.75; b. 23.38; c. 11.69; d. 5.84

Sun

1. (0,1)
2. In exponential decay models, the starting value is decreasing by a constant multiplier. Although the numbers keep getting smaller, they will never reach zero.
3. $f(0)$ means the starting value of the car; $f(-1)$ means what the car was worth the previous year.
4. $y = 500(.75)^2$; 281.25 mg
5. Answers will vary. A traditional bouncy ball can be modeled with a decay function. The function in this problem is a growth model. One possible scenario: There are 80 bacteria in an initial population and it is increasing by 25% every hour.
6. Answers will vary. a. $600 is deposited in an account earning 5% interest; b. half-life or radioactive material starts with 40 kg and decreases by 50% each time period.

Star

1. In exponential decay models, the starting value is decreasing by a constant multiplier. Although the numbers keep getting smaller, they will never reach zero.
2. $f(0)$ means the starting value of the car; $f(-1)$ means what the car was worth the previous year.
3. Answers will vary. One possible solution: A table was purchased for $600 and is depreciating at a rate of 25% per year. How much will it be worth after t years?
4. The first scenario is linear: $y = 45,000 + 2,000x$. The second scenario is exponential: $y = 45,000(1 + .04)^x$. The first plan results in a higher salary for the first 6 years. After that point, the second plan results in a higher salary.
5. $5 < y \le 5,120$
6. To change from annual interest to monthly, the interest amount must be divided by 12 to determine the amount of interest per month. The exponent t would also have to be divided by 12. Insert the function into the calculator, and then look for the amount in the y-column after 1.5 years (which is 18 months).

Lesson 3: Exponential Decay: Exit Slip

1. $y = A(1 - r)^x$; y = final value, A = initial or starting value; r = rate of decay; x = time period
2. Answers will vary. Exponential decay models are used for items that are declining in value by the same percentage each period.

Lesson 4: Multiplication Properties of Exponents: Agenda

Imperatives

1. a. a^2; b. b^3; c. c^4; d. $2a^2$; e. $4c^2d$; f. 3^4
2. a. $2 \cdot 2 \cdot 2 \cdot 2 \cdot 2$; b. $b \cdot b \cdot b \cdot b \cdot b \cdot b \cdot b$; c. $a \cdot a \cdot a \cdot a \cdot a \cdot a \cdot a$; c. $4 \cdot 4 \cdot 4 \cdot 4 \cdot 4 \cdot 2 \cdot 2 \cdot 2$
3. a. x^7; b. $6^5 = 7,776$; c. $5^{11} = 48,828,125$; d. a^8b^2; e. b; f. $4 \cdot 2^7 = 512$; g. a^5x^3; h. $4^5y = 1,024y$
4. a. x^{12}; b. $4^6 = 4,096$; c. $2^{10} = 1,024$; d. $27x^3$; e. $64x^6$; f. $729b^9$; g. $729x^6$; h. $1,024b^4$

Negotiables

1. Answers will vary. When the negative is on the outside of the parentheses, it must be applied at the very end of the simplification process. $-(2x)^2 = -4x^2$ while $(-2x)^2 = 4x^2$

2. a. Answers will vary. The expression $x^2 \cdot t^2$ cannot be simplified. The multiplication rule of exponents cannot be used because the bases are not the same. b. Answers will vary. The expressions $(x^2)^3$ and $x^2 \cdot x^3$ are not equivalent expressions. The power to a power rule simplifies the first expression to x^6, and the second expression simplifies to x^5.

3. $V = 12x^5$; Students should draw a three-dimensional storage box with the required dimensions.

Options

1. Answers will vary. $(a+1)^3$ means that you have to multiply $a+1$ by itself three times. It simplifies to $a^3 + 3a^2 + 3a + 1$, not $a^3 + 1^3$.

Lesson 4: Multiplication Properties of Exponents: Hexagon Puzzles

Moon	Sun	Star
A→ a^2b^2	A→ a^2b^2	A→ $10x^2$
B→ $2 \cdot 2 \cdot 2 \cdot 2 \cdot 2 \cdot 2 \cdot 2 \cdot 2$	B→ $2 \cdot 2 \cdot 2 \cdot 2 \cdot 2 \cdot 2 \cdot 2 \cdot 2$	B→ $-3x^5$
C→ b^2c^2	C→ b^2c^2	C→ $2x^{12}$
D→ $t \cdot t \cdot r \cdot r \cdot r \cdot r$	D→ $t \cdot t \cdot t \cdot r \cdot r \cdot r \cdot r \cdot r$	D→ $-8t^4x^6$
E→ x^5	E→ x^8	E→ x
F→ x^4	F→ x^4	F→ xt^8
G→ $2x^7$	G→ x^{12}	G→ $8a^5b^3$
H→ t^{10}	H→ t^{10}	H→ $-49{,}152x^{-2}$ or $\dfrac{-49{,}152}{x^2}$
I→ x^7	I→ x	I→ x^8
J→ x^5t^7	J→ xt^8	J→ $32d^{20}$
K→ $8a^5b^7$	K→ $8a^5b^3$	K→ $4{,}096$
L→ $6{,}144x^2$	L→ $-49{,}152x^{-2}$ or $\dfrac{-49{,}152}{x^2}$	L→ $81x^{12}$
M→ x^4	M→ x^6	M→ $4x^{10}$
N→ b^6	N→ d^{12}	N→ $\dfrac{c^{18}}{216}$
O→ x^6	O→ $262{,}144$	O→ $32x^{25}$
P→ e^8	P→ $81x^{12}$	P→ $64b^8$
Q→ $262{,}144$	Q→ x^{10}	Q→ $2x^{18}$
R→ $81x^{12}$	R→ c^{18}	R→ $81c^4x^{11}$
S→ x^{10}	S→ $8x^{21}$	S→ $-28c^{11}$
T→ d^{21}	T→ $512b^3$	T→ $18c^5$
U→ $8x^{21}$	U→ x^{23}	U→ $-4x^8$

Moon	Sun	Star
V→$64b^{12}$	V→$c^4 x^{11}$	V→$-16x^{10}$
W→$c^4 x^8$	W→$28c^{17}$	W→$12x^{10}$
X→x^{21}	X→$18c^5$	X→$-36x^{10}$

Lesson 4: Multiplication Properties of Exponents: Exit Slip

1. Answers will vary. a. When multiplying exponents with the same base, the exponents are added: $a^m \cdot a^n$ simplifies to a^{m+n}; b. When raising a power to a power, the exponents are multiplied: $(a^m)^n$ simplifies to a^{mn}.

2. Answers will vary. Possible examples: $b^4 \cdot b^2 = b^6$ and $x^5 \cdot x^{12} = x^{17}$; $(b^2)^4 = b^8$ and $(x^5)^4 = x^{20}$

Lesson 5: Zero and Negative Exponents: Agenda

Imperatives

1.

a.

2^{-4}	2^{-3}	2^{-2}	2^{-1}	2^0	2^1	2^2	2^3	2^4	2^5
$\frac{1}{16}$	$\frac{1}{8}$	$\frac{1}{4}$	$\frac{1}{2}$	1	2	4	8	16	32

b.

4^{-4}	4^{-3}	4^{-2}	4^{-1}	4^0	4^1	4^2	4^3	4^4	4^5
$\frac{1}{256}$	$\frac{1}{64}$	$\frac{1}{16}$	$\frac{1}{4}$	1	4	16	64	256	1,024

2. a. 1; b. 1; c. 5; d. 1; e. 1; f. $\frac{1}{5}$

3. a. $\frac{1}{16}$; b. $\frac{1}{16}$; c. $\frac{1}{2}$; d. $\frac{1}{4}$; e. 27; f. $\frac{1}{2,048}$; g. 1; h. 1

4. a. $\frac{1}{x^2}$; b. $\frac{1}{y^4}$; c. $\frac{4}{x^3}$; d. $\frac{2}{a^3}$; e. $2t^3$; f. $\frac{1}{6m^3}$; g. $\frac{x^4}{y^6}$; h. 1

Negotiables

1. Answers will vary. $2a^{-3}$ is not equivalent to $\frac{1}{2a^3}$. $2a^{-3}$ simplifies to $\frac{2}{a^3}$ because only the a is being raised to the -3 power.

2. a. $\frac{2}{a^4}$; b. $\frac{-1}{81}$; c. $\frac{-5}{x^3}$; d. $\frac{1}{25a^2}$

3. Answers will vary. 2^x is an exponential growth model whereas $\left(\frac{1}{2}\right)^x$ is an exponential decay model. Both are exponential functions because of the variable in the exponent.

Options

1. a. $y = 18,233(1+0.05)^{-4}$, the original amount deposited was approximately $15,000; b. Answers will vary. Negative exponents can be used to look backward in time.

Lesson 5: Zero and Negative Exponents: Tic-Tac-Toe Board

1. a. $5^1 = 5, 5^0 = 1, 5^{-1} = \frac{1}{5}$; b. $-5^1 = -5, -5^0 = 1, -5^{-1} = -\frac{1}{5}$; c. $(-5)^1 = -5, (-5)^0 = 1, (-5)^{-1} = -\frac{1}{5}$

2. a. $x = -1$; b. $x = 4$; c. $x = 36$

3. Answers will vary. One solution: $f(t)$ is the amount in an account at any time t earning interest of 3% with an initial deposit of $6,000. At $t = 0$, $f(t) = 6,000$ because this is time zero and any number or variable raised to the zero power is 1.

4. Answers will vary depending on what year "now" is (in 2012, it is worth approximately $29,228); $y = 20,500(1+0.03)^x$. In 2007, the antique car was worth approximately $25,212.

5. $10,000 = A(1+0.03)^x$; approximately $8,131

6. a. $5b^1 = 5b, 5b^0 = 5, 5b^{-1} = \frac{5}{b}$; b. $-5(1)^3 = -5$, $-5(0)^3 = 0$, $-5(-1)^3 = 5$; c. $(-5b)^1 = -5b$, $(-5b)^0 = 1$, $(-5b)^{-1} = \frac{-1}{5b}$

7. a. $2^{-2} = \frac{1}{4}$; b. $(-2)^{-3} = -\frac{1}{8}$; c. $3^{-2} = \frac{1}{9}$; d. $(-2)^0 = 1$

8. The car was worth approximately $13,207 three years earlier and approximately $22,022 seven years ago.

9. a. a^3c^2; b. $\frac{1}{c^2d^3}$; c. $\frac{a^2}{b^2}$

Lesson 5: Zero and Negative Exponents: Exit Slip

1. Answers will vary. a. Any number or variable raised to the zero power is 1: $a^0 = 1$; b. A number raised to a negative power becomes 1 divided by that number raised to the same exponent but now positive: a^{-1} simplifies to $\frac{1}{a^1}$.

2. Answers will vary. Examples could include: $b^0 = 1$ and $4^0 = 1$; $b^{-3} = \frac{1}{b^3}$ and $2^{-4} = \frac{1}{16}$.

Lesson 6: Division Properties of Exponents: Agenda

Imperatives
1. a. $a \cdot a \cdot a$; b. 3; c. $\frac{1}{6 \cdot 6}$; d. $a \cdot c$

2. a. $\frac{27}{64}$; b. $\frac{81}{y^4}$; c. $\frac{81}{16}$; d. $\frac{1}{4}$

3. a. x^2; b. $\frac{x^2}{4y^2}$; c. $\frac{125}{27}$; d. 1; e. 1; f. $\frac{x^4}{4y^6}$; g. $32x^5$; h. x^2y^4

Negotiables
1. a. xy; b. $\frac{1}{xt}$; c. $\frac{y^3}{z^4}$; d. $\frac{y^3}{x^4z^2}$

2. a. 5; b. 2; c. -8; d. 6

3. a. simplifies to y^2z^3 (negative exponents must be simplified); b. it is simplified because there are no negative exponents

Options

1. $\dfrac{-108}{x^7 y^{22}}$

2. $\dfrac{-9}{x^{40} y}$

Lesson 6: Division Properties of Exponents: Hexagon Puzzles

Moon	Sun	Star
A→a^4	A→$\dfrac{1}{a^3}$	A→1
B→a^4	B→$\dfrac{1}{64}$	B→$\dfrac{1}{a^2}$
C→5	C→$\dfrac{1}{16}$	C→$\dfrac{1}{a^{12}}$
D→t^2	D→$\dfrac{1}{y^5}$	D→a^{12}
E→$\dfrac{1}{a^3}$	E→1	E→$\dfrac{y^8}{x^4}$
F→$\dfrac{1}{64}$	F→$\dfrac{1}{a^2}$	F→$\dfrac{t^6}{4x^6}$
G→$\dfrac{1}{16}$	G→$\dfrac{1}{a^{12}}$	G→$\dfrac{-1}{2t^2 w^2}$
H→$\dfrac{1}{y^5}$	H→a^{12}	H→$\dfrac{w^2}{2t^3 z}$
I→x^2	I→$\dfrac{y^8}{x^4}$	I→4
J→1	J→$\dfrac{t^6}{4x^6}$	J→−10
K→$\dfrac{1}{a^2}$	K→$\dfrac{1}{-2t^2 w^2}$	K→25
L→$\dfrac{1}{a^{12}}$	L→$\dfrac{w^2}{2t^3 z}$	L→3
M→$\dfrac{1}{x^4}$	M→$\dfrac{-x^4}{2}$	M→1
N→$\dfrac{y^8}{x^4}$	N→$-2x^{10}$	N→−1
O→$\dfrac{t^6}{4x^6}$	O→$\dfrac{3}{t^8}$	O→9 and 3

Moon	Sun	Star
P→$\dfrac{1}{-2t^2w^2}$	P→$\dfrac{-1}{2t^3}$	P→2
Q→$\dfrac{-x^4}{2}$	Q→4	Q→3 and 4
R→$-2x^{10}$	R→−7	R→2 and 6
S→$\dfrac{3}{t^8}$	S→25	S→$x^{10}z^{10}$
T→$\dfrac{-1}{2t^3}$	T→3	T→2 and 36
U→4	U→1	U→−5
V→3	V→−1	V→−2 and −2
W→25	W→9 and 3	W→−1
X→3	X→2	X→3

Lesson 6: Division Properties of Exponents: Exit Slip

1. Answers will vary. a. $\dfrac{a^m}{a^n}$ simplifies to a^{m-n}; when dividing exponents with the same bases, subtract the exponents; b. $\left(\dfrac{a}{b}\right)^m$ simplifies to $\dfrac{a^m}{b^m}$; fractions raised to a power become the numerator raised to the power divided by the denominator raised to the same power.

2. Answers will vary. Possible solutions: $\dfrac{2^6}{2^4}$ becomes 2^2 or 4 and $\dfrac{x^5}{x^{-4}}$ becomes x^9; $\left(\dfrac{2}{b}\right)^3$ becomes $\dfrac{2^3}{b^3}$ or $\dfrac{8}{b^3}$ and $\left(\dfrac{x^5}{y^2}\right)^2$ becomes $\dfrac{x^{10}}{y^4}$.

Unit 4: Quadratic Functions

Quadratic Models: Preassessment

1.

vertex $(-1,0)$; axis of symmetry: $x=-1$; minimum: $(-1,0)$

Unit 4: Quadratic Functions

2. a. quadratic function; b. linear function

3. $(2x+1)(3x-2)$

4. $\dfrac{-3\pm\sqrt{41}}{-4}$

5. $x=-5$ and $x=-1$

6. Answers will vary. Possible answers: a. x^2+6x+2; b. x^2-2x+1; c. x^2-2x+2

7. a. 5 ft; b. $h(0)=5$ represents the ball's height at time = 0, when the ball is 5 ft off the ground ready to be thrown; c. 71 ft.; d. The ball is 30 feet above the ground when $x=.39$ seconds and $x=3.98$ seconds; e. 81.56 ft; f. 4.45 seconds

Lesson 1: Introduction: Exit Slip

1. Plug the x-values into the equation and solve for y; $(-3,2)(-2,-1)(-1,-2)(0,-1)(1,2)$

2. The graph will contain the points: $(-3,2)(-2,-1)(-1,-2)(0,-1)(1,2)$; the shape of the graph is a parabola

3. $(-1,-2)$

4. $x=-1$

Lesson 2: Characteristics of Quadratics: The Handshake Challenge

1. Answers will vary. Possible solutions: experiment with group and look for a pattern or set up a table.

2. $1, 3, 6, 10, 15, 21, 28, 36$

3. 10 people = 45; 11 people = 55

4. Common formula: total handshakes $= 0.5x^2-0.5x$; the equation can be derived using the quadratic regression function on the graphing calculator using the data collected from the handshake experiment.

Extension Questions

1. 66 matches

2. $1, 3, 6, 10, 15$; $-.5n^2-0.5n$, where n is the picture number plus 1

Lesson 2: Characteristics of Quadratics: Agenda

Imperatives

1. a. $a=1$, $b=2$, $c=-5$; b. $a=-2$, $b=-4$, $c=6$; c. $a=-1$, $b=2$, $c=0$; d. $a=5$, $b=0$, $c=2$

2. a. $x=1$; b. $x=1$; c. $x=3$; d. $x=\dfrac{5}{2}$

3. a. $(1,3)$; b. $(1,2)$; c. $(2,4)$

Negotiables

1. Vertex is $(0,-2)$, line of symmetry is $x=0$, minimum

2. Vertex is $(-1,-1)$, line of symmetry is $x=-1$, minimum

3. Vertex is $(-0.5,-5.75)$, line of symmetry is $x=-0.5$, maximum

4. Vertex is $(1,-4)$, line of symmetry is $x=1$, minimum

5. Vertex is $(-2,4)$, line of symmetry is $x=-2$, maximum

6. Vertex is $(0,-9)$, line of symmetry is $x=0$, minimum

Options

1. a. 70; b. 2.5; c. 51.5; d. 4.4

Lesson 2: Characteristics of Quadratics: Tic-Tac-Toe Board

1. a. linear; b. quadratic; c. linear; d. exponential; e. quadratic
2. a. opens up with 1 as the coefficient; b. opens down with -5 as the coefficient; c. opens up with 3 as the coefficient; d. opens up with $\dfrac{3}{2}$ as the coefficient
3. Answers will vary. Possible answers: a. $y = x^2 + 2x + 1$; b. $-x^2 + 2x + 1$; c. $y = x^2 + 1$; d. $y = -x^2 + 2x + 5$
4. The second differences are constant.
5. a. $(-1.5, -4.25)$ and $x = -1.5$; b. $(-.75, 1.125)$ and $x = -.75$; c. $(-.75, -1.625)$ and $x = -.75$
6. a. $(-2, 0)$ and $(1, 0)$; b. $(-3, 0)$ and $(2, 0)$
7. Answers will vary. Graphs: a. opens upward, crosses the y-axis at 2, and is wider than the second quadratic; b. opens upward, crosses the y-axis at 6, and is narrower than the other quadratics; c. opens downward, crosses the y-axis at 2, is wider than the second quadratic, and is a mirror image of the first quadratic.
8. a. $(0, 0)$ and $(2, 0)$; b. $(1, 0)$
9. 4

Lesson 2: Characteristics of Quadratics: Forms of Quadratic Equations

Equation	Form	Starting Value	Vertex	Axis of Symmetry	Zeros
$y = 2(x+2)^2 - 4$	vertex form	—	$(-2, -4)$	$x = -2$	$-3.41, -0.59$
$y = 2(x-2)^2 + 5$	vertex form	—	$(2, 5)$	$x = 2$	none
$y = 3(x-2)(x-4)$	factored	—	$(3, -3)$	$x = 3$	$2, 4$
$y = -6(x-3)^2 + 6$	vertex form	—	$(3, 6)$	$x = 3$	$2, 4$
$y = 2x^2 + 4x - 6$	standard	-6	$(-1, -8)$	$x = -1$	$-3, 1$
$y = x^2 - 3x + 7$	standard	7	$(1.5, 4.75)$	$x = 1.5$	none

Lesson 2: Characteristics of Quadratics: Hexagon Puzzles

Moon	Sun	Star
A→3	A→−1	A→Answers will vary. Possible answer: $y = x^2 + 1$
B→$(2, -3)$	B→$(-5, -1)$	B→ $y = 2x^2 - 2x + 3$
C→−8	C→ $y = x^2 + x + 6$	C→ $y = 2x^2 + x + 6$
D→2 and −7	D→−3 and 2	D→ $y = (x-0)^2 + 4$
E→1 and −4	E →Vertex	E→ $x = 2$

Moon	Sun	Star
F→Answers will vary. Possible answer: $y = -2x^2 - 2x + 3$	F→Answers will vary. Possible answer: $y = 2x^2 - 2x + 3$	F→ $x = -3$
G→$(-3, 1)$	G→Zeros	G→ $x = -3$
H→ $y = 2(x-3)^2 - 1$	H→ $y = 2(x+2)^2 + 8$	H→Answers will vary. Possible answer: $y = 2(x+7)^2 - 3$
I→y-intercept	I→Answers will vary. Possible answer: $y = x^2 + 1$	I→ Answers will vary. Possible answer: $y = (x-1)^2 + 5$ and $y = -2(x-1)^2 + 5$
J→Vertex	J→ $y = (x-0)^2 + 4$	J→Answers will vary. Possible answer: $y = (x+2)(x-1)$ and $y = 2(x+2)(x-1)$
K→Zeros	K→Answers will vary. Possible answer: $y = (x-3)^2 + 1$	K→Answers will vary. Possible answer: $y = (x+1)^2 + 1$
L→Starting value	L→Answers will vary. Possible answer: $y = x^2 + 2x$	L→Answers will vary. Possible answer: $y = 2x^2 - 3$
M→ $x = 2$	M→ $x = 0$	M→ $y = x^2 + 16x + 62$
N→ $x = -1$	N→ $x = -3$	N→ $y = 2x^2 - 12x + 20$
O→ $x = -2$	O→ $x = 5$	O→ $y = -3x^2 + 6x - 1$
P→ $x = 1$	P→ $x = 0.5$	P→ $x = 0.5$
Q→1 and -4	Q→Answers will vary. Possible answer: $y = (x-1)^2 + 3$ and $y = 2(x-1)^2 + 3$	Q→ $y = x^2 - 5$
R→-5 and -2	R→Answers will vary. Possible answer: $y = (x-3)(x-6)$ and $y = 2(x-3)(x-6)$	R→ $y = (x-1)^2 - 2$
S→3 and -1	S→Answers will vary. Possible answer: $y = (x+3)^2 + 2$ and $y = 2(x+3)^2 + 2$	S→ $y = x^2 + 3$
T→-3 and -1	T→Answers will vary. Possible answer: $y = (x+2)(x+9)$ and $y = 2(x+2)(x+9)$	T→ $y = (x+1)^2 - 8$
U→Answers will vary. Possible answer: $y = (x-1)^2 + 3$	U→ $y = x^2 + 6x + 10$	U→Up 4 and to the left 1
V→ $y = (x-3)(x-6)$	V→ $y = 2x^2 - 8x + 9$	V→ $y = (x+3)^2 - 5$
W→ $y = (x+3)^2 - 2$	W→ $y = 2x^2 - 12x + 20$	W→Narrower and down 2
X→ $y = (x+2)(x+9)$	X→ $y = -3x^2 + 6x - 1$	X→Because the x-value is being squared and any negative number squared is positive

Lesson 2: Characteristics of Quadratics: Exit Slip

1. If $a > 0$, the graph opens upward; if $a < 0$, the graph opens downward
2. c is the y-intercept; when c changes, the graph shifts up or down
3. Use $x = \dfrac{-b}{2a}$ to find the x-value of the vertex. Substitute the x-value into the equation to find the y-value. Using the equation given, $a = 2$, $b = -3$, and $c = 1$. $x = \dfrac{-(-3)}{2 \cdot 2}$, $x = \dfrac{3}{4}$. Substituting $\dfrac{3}{4}$ or 0.75 into the equation $y = 2(0.75)^2 - 3(0.75) + 1$ results in a y-value of -0.125. The vertex is $(0.75, -0.125)$.
4. The axis of symmetry is the line that divides the parabola into mirror images. It is the x-value of the vertex.

Lesson 3: Solving Using the Quadratic Formula: Agenda

Imperatives

1. First change the equation so that it is in standard form by subtracting the 7 from both sides of the equation. The equation will then read $0 = 2x^2 - 4x - 6$. Then identify the a, b, and c values to plug into the quadratic formula.
2. a. $2x^2 + 3x - 4 = 0$; b. $0 = -2x^2 - 6x - 5$; c. $-4x^2 - x + 2 = 0$; d. $-5x^2 + x + 6 = 0$
3. a. $x = 3$ and 0.25; b. $x = 5$ and -3; c. $x = 2$ and 1; d. $x = -2$ and 4

Negotiables

1. Approximately 3.95 seconds using $h(t) = -16t^2 + 250$. There are two times when the height of the sunglasses is 0. One of these times is negative, so this is the answer that can be eliminated given the context of the story problem.
2. 3.79 seconds using $h(t) = -16t^2 + 60t + 2.5$

Options

1. These instructions are for using a Texas Instruments graphing calculator. The data can be loaded into a table in the calculator and an equation can be created to model the data. First, press the STAT key then select EDIT. Load the x- and y-values into two lists, being careful that the data are properly paired and that there are the same number of data in each list. Note the list numbers where the data are located. Select STAT, CALC, QuadReg. If the data is stored in L1 and L2, type these into the calculator as L1, L2 and then press enter. The calculator will identify the a, b, and c values to plug into the standard form of a quadratic equation. This equation can then be loaded into the $y =$ function on the calculator, and the table can be checked for the values that were given in the problem. The equation that models the given data is $y = -2x^2 - 5x + 2$.

Lesson 3: Solving Using the Quadratic Formula: Think Dots

Moon

1. Vertex is the coordinate point of the lowest or highest point of the parabola; minimum value is the vertex if the parabola opens upward; maximum value is the vertex if the parabola opens downward; axis of symmetry is the $x =$ line that cuts the parabola into mirror images. It is also the x-value of the vertex.

2. The graph of a quadratic with two solutions is a parabola that has two x-intercepts. The graph of a quadratic with one solution is a parabola that has just one x-intercept. The graph of a quadratic with zero solutions has no x-intercept.

3. Answers will vary. Possible solution: $y = -2x^2 + 3x + 1$

4. Quadratic formula: $\dfrac{-b \pm \sqrt{b^2 - 4ac}}{2a}$. In order to use the quadratic formula, first identify the a, b, and c values from the equation. To find the x-values or the solutions, plug the a, b, and c values into the formula. Take the opposite of b and then add and subtract the square root of b squared minus 4 times the a-value times the c-value. This quantity is then divided by 2 times the a-value. There will either be two, one, or no real solutions.

5. a. $a = 1$, $b = 2$, $c = 5$; b. $a = -4$, $b = -3$, $c = 1$; c. $a = -2$, $b = 1$, $c = 9$; d. $a = -1$, $b = 5$, $c = 4$

6. $a = 1$, $b = 9$, $c = 14$; solutions are $x = -2$ and $x = -7$. Graph the equation to check the solutions using the calculator. Plug in -2 and -7 for x in the equation.

Sun

1. Vertex is the coordinate point of the lowest or highest point of the parabola; minimum value is the vertex if the parabola opens upward; maximum value is the vertex if the parabola opens downward; axis of symmetry is the $x =$ line that cuts the parabola into mirror images. It is also the x-value of the vertex.

2. a. $a = 3$, $b = -2$, $c = 4$; b. $a = -4$, $b = -3$, $c = 0$; c. $a = -4$, $b = 2$, $c = -3$; d. $a = -6$, $b = -1$, $c = 6$

3. Rewrite the equation into standard form. $a = 1$, $b = 9$, $c = 14$; solutions are $x = -2$ and $x = -7$. Graph the equation to check the solutions using the calculator. Plug in -2 and -7 for x in the equation.

4. No real solutions. After using the quadratic formula, the answer includes a negative square root. Graphing this quadratic on the calculator will result in a parabola that does not cross the x-axis.

5. a. 200 ft/sec; b. 100 ft

6. Approximately 0.55 seconds and -1.80 seconds. The second answer can be eliminated because it represents negative time.

Star

1. Vertex is the coordinate point of the lowest or highest point of the parabola; minimum value is the vertex if the parabola opens upward; maximum value is the vertex if the parabola opens downward; axis of symmetry is the $x =$ line that cuts the parabola into mirror images. It is also the x-value of the vertex.

2. No real solutions. Graphing this quadratic on the calculator will result in a parabola that does not cross the x-axis.

3. Programs can be found online.

4. $h(t) = -16t^2 - 100t + 200$. Approximately -7.84 seconds and 1.59 seconds. The first answer can be eliminated because it represents negative time.

5. Answers will vary. Possible solution: If t = time in seconds and $h(t)$ represents height off the ground in feet, several ordered pairs could be $(0, 30)(0.1, 32)(0.2, 29)(0.3, 23)$. The equation that models these data is $h(t) = -200t^2 + 36t + 30.1$.

6. Answers will vary. Possible solution: A ball is thrown from a height of 58 feet at an initial speed of 60 ft/sec. How long does it take for the ball to hit the ground?

Lesson 3: Solving Using the Quadratic Formula: Exit Slip

1. $\dfrac{-b \pm \sqrt{b^2 - 4ac}}{2a}$

2. The quadratic formula can be used to solve any quadratic because any quadratic can be written in standard form. Not all quadratic equations can be solved using factoring and completing the square is often difficult.

3. One of the two solutions may be a negative answer, which would represent negative time. This is generally the solution that can be eliminated.

Lesson 4: Solving by Factoring: Agenda $x^2 + bx + c$

Imperatives

1. Factors of c that add to b

2. The first equation goes with the third factorization; the second equation goes with the second factorization; the third equation goes with the fourth factorization; the fourth equation goes with the first factorization.

3. a. $(x+4)(x+3)$; b. $(t+2)(t+6)$; c. $(m-5)(m+2)$; d. $(b-5)(b+4)$; e. $(x-6)(x+3)$; f. $(n+8)(n-3)$; g. $(x+7)(x-40)$

Negotiables

1. a. $x = -12$ and $x = -7$; b. $b = 2$ and $b = -4$; c. $x = 2$ and $x = 15$; d. $n = 19$ and $n = 1$; e. $z = -25$ and $z = -40$; f. $x = -15$ and $x = -1$

2. a. $(x-10)$; b. $x = 20$ and $x = -5$ (the negative answer can be eliminated); c. 15 by 10

Options

1. Based on the expression $x^2 + 3x - 40$, the factors of -40 will have to include one positive and one negative factor.

Lesson 4: Solving by Factoring: Agenda $ax^2 + bx + c$

Imperatives

1. Factors of $a \cdot c$ that add to b

2. The first equation goes with the third factorization; the second equation goes with the first factorization; the third equation goes with the fourth factorization; the fourth equation goes with the second factorization.

3. a. $(3x+1)(x-6)$; b. $(t+5)(3t+1)$; c. $(5m+1)(m-2)$; d. $(b-3)(4b-14)$; e. $(2x-1)(4x+3)$; f. $(n-2)(6n+1)$; g. $(2x-1)(x+10)$

Negotiables

1. a. $x = \dfrac{-3}{2}$ and $x = 7$; b. $b = \dfrac{5}{4}$ and $b = \dfrac{1}{2}$; c. $x = \dfrac{4}{3}$ and $x = 11$; d. $n = \dfrac{1}{2}$ and $n = \dfrac{4}{7}$; e. $z = \dfrac{-1}{3}$ and $z = -11$; f. $x = \dfrac{1}{6}$ and $x = -5$

2. length = 10 and width = 18

Options

1. $h(0) = 0$ is when the ball is on the ground before it is kicked; b. 2 times; c. 25 feet; d. 1.25 seconds

Lesson 4: Solving by Factoring: Tic-Tac-Toe Board

1. a. cannot be factored; b. can be factored; c. cannot be factored; d. can be factored
2. a. $h(t) =$ the height at time t, $t =$ time, and $s =$ initial height; b. $h(t) =$ height at time t, $v =$ initial velocity, $t =$ time, and $s =$ initial height
3. Approximately 2.19 seconds and the ball is 82.56 ft in the air
4. a. $(3t - 2)(t + 3)$; b. $(3x - 4)(x + 2)$; c. not factorable; d. not factorable; e. not factorable
5. a. 8 feet; b. 1 second
6. a. not factorable; b. can be factored; c. not factorable; d. can be factored
7. Answers will vary. a. Possible solution: $h(x) = (2x + 1)(5x - 3)$; b. Possible solution: $f(x) = (3x - 4)(3x - 2)$
8. a. $(3x - 1)(5x - 2)$ where $x = \dfrac{1}{3}$ and $x = \dfrac{2}{5}$; b. $(4x - 2)(3x + 1)$ where $x = \dfrac{1}{2}$ and $x = \dfrac{-1}{3}$
9. The area of the large rectangle that includes the path and the garden can be represented by $(x + 4)(x + 10)$, and this area has to be the 40 square feet of the garden plus the 51 square feet for the path. So, $(x + 4)(x + 10) = 91$. Multiply and then subtract 91 from both sides and you get $x^2 + 14x - 51 = 0$. This now factors to $(x + 17)(x - 3) = 0$. Using the zero product property, either $(x + 17) = 0$ or $(x - 3) = 0$. Solving each of these for x results in $x = -17$ or $x = 3$. The negative answer can be eliminated. The path should be 3 feet wide.

Lesson 4: Solving by Factoring: Exit Slip

1. Solving a quadratic means finding the zeros or where the quadratic intersects the x-axis.
2. Quadratics need to be in the form of $0 = ax^2 + bx + c$ in order to factor. Subtract 6 from both sides.
3. Answers will vary. Looking for factors of -6 that add to -5. Factors of 6 are 1, 2, 3, 6. One factor will have to be positive and one will have to be negative. Because of the small numbers, guess and check can result in $(2x + 1)(x - 3)$.

Lesson 5: Solving by Completing the Square: Agenda $x^2 + bx = c$

Imperatives

1. a. $x = 3$ and $x = -3$; b. $x = 6$ and $x = -6$; c. $x = \pm\sqrt{18}$; d. $x = \pm\sqrt{10}$; e. $x = 9$ and $x = -9$; f. no real solutions
2. The first trinomial goes with the third squared binomial; the second trinomial goes with the first squared binomial; the third trinomial goes with the second squared binomial; the fourth binomial goes with the fourth squared binomial.
3. Six steps: a. move variables to one side of the equation and constants to the other side; b. divide every term by the coefficient of x^2; c. divide b by 2 and square the result, and add the squared number to both sides of the equation; d. write the variable side of the equation as a perfect square binomial, and simplify the constant-term side; e. take the square root of both sides of the equation; f. solve for x.
4. a. add 4 to both sides; b. add 25 to both sides; c. add 9 to both sides; d. add 16 to both sides

Negotiables

1. Answers will vary. Possible solutions: a. In order to solve a quadratic equation using completing the square, the b-value is divided by 2. When you divide an even number by 2, you get an integer and not a decimal. b. The a-value must be a one, so you would have to divide all of the terms in the quadratic equation by 5.

2. Solutions: $x = -6$ and $x = 2$

Options

1. a. factoring because it can be factored; b. quadratic formula because it cannot be factored and completing the square would not be easy; c. graphing because it cannot be factored; d. completing the square because $a = 1$ and b is an even number.

Lesson 5: Solving by Completing the Square: Think Dots

Moon

1. a. $x^2 - 1$; b. $x^2 - 6x + 9$; c. $x^2 + 6x + 9$; d. $x^2 - 10x + 25$
2. a. $(x + 4)^2$; b. $(x - 3)^2$; c. $(x + 5)^2$
3. a. $x = \pm 5$; b. $x = \pm 7$; c. $x = \pm 10$; d. $x = \pm\sqrt{15}$
4. a. 1; b. 9; c. 36; d. 4
5. a. $x = -8$ and $x = 0$; b. $x = 5$ and $x = -1$
6. $P(0) = \$50$ represents the starting price of the stock; $74 after 6 weeks; $(1, 49)$ is the vertex, and it represents the lowest price of the stock, which occurred in Week 1.

Sun

1. a. $(x - 5)^2$; b. $(x - 9)^2$; c. $(2x - 3)^2$
2. a. $x = -6$ and $x = 4$; b. $x = -6$ and $x = 8$; c. $x = -8$ and $x = 2$
3. a. $x = \pm 5$; b. $x = \pm 7$; c. $x = \pm 6$
4. a. 9; b. 1; c. 25; d. 16
5. a. $x = -4 \pm \sqrt{10}$; b. $x = 1 \pm \sqrt{5}$
6. a. 100 ft after 2.5 seconds; b. $t = 5$ and $t = 0$ are the times the firework is on the ground

Star

1. Answers will vary. Possible solution: $x^2 + 4x - 6 = 0$ or $3x^2 - 21x + 10 = -5$
2. Equation: $3x(x + 8) = 540$. Solutions: $x = 10$ and $x = -18$ but only $x = 10$ is a solution because there cannot be negative dimensions. The rug is 30 ft by 18 ft.
3. Answers will vary. Possible solution: a. The discriminant is a negative number. When it is graphed, the parabola does not intersect the x-axis; b. $2x^2 + 5x + 10 = 0$
4. a. $x = 6$ and $x = -4$; b. $x = -2 \pm \sqrt{23}$; c. $x = -5 \pm \sqrt{7}$
5. a. $x = -2 \pm \sqrt{2}$; b. $x = 2 \pm \sqrt{8}$
6. Answers will vary. Possible solution: This equation models an object that is thrown or launched from an initial height of 0 feet at an initial velocity of 60 ft/sec. Solutions are $x = 0$ and $x = 3.75$, which means the object is on the ground at $t = 0$ and then hits the ground when $t = 3.75$.

Lesson 5: Solving by Completing the Square: Exit Slip

1. a. move variables to one side of the equation and constants to the other side (add 8 to both sides); b. divide each term by the coefficient of x^2 (divide all terms by 2); c. divide the b-value by 2 and then square the result, and add the squared number to both sides of the equation (add 1 to both sides of the equation); d. write the variable side of the equation as a perfect square binomial $(x-1)^2$, and simplify the side with the constants (5); e. take the square root of both sides of the equation; f. solve for x. Solutions: $x = 1 \pm \sqrt{5}$

References

Benson, B. (1997). Coming to terms: Scaffolding. *The English Journal*, 86(7), 126–127.

Common Core State Standards Initiative. (n.d.). *About the standards*. Retrieved from http://www.corestandards.org/about-the-standards

Demi. (1997). *One grain of rice*. New York, NY: Scholastic Press.

Ellis, E. L. (2000). *The LINCS vocabulary strategy* (Rev. ed.). Lawrence, KS: Edge Enterprises.

Johnson, D. W., Johnson, R. T., & Smith, K. A. (2000). Constructive controversy. *Change, 32*(1), 28.

National Council of Teachers of Mathematics. (2010). *NCTM supports teachers and administrators to implement Common Core Standards*. Retrieved from http://www.nctm.org/news/content.aspx?id=26083

Tomlinson, C. (2001). *How to differentiate instruction in mixed-ability classrooms*. Alexandria, VA: Association for Supervision and Curriculum Development.

Santa, C. (1988). *Content reading including study systems*. Dubuque, IA: Kendall/Hunt.

Sternberg, R. (2010). Assessment of gifted students for identification purposes: New techniques for a new millennium. *Learning and Individual Differences, 20*, 327–336.

About the Author

Kelli Jurek has been teaching mathematics for the last 7 years after changing careers to become a professional educator. Like most educators, she has worked in multicultural settings, and her classrooms include students with different learning styles, varied levels of motivation, and unique interests. She recently earned a Master of Educational Differentiation degree from Grand Valley State University. She lives in Rockford, MI, with her husband, Kevin, and several pets. She has two grown daughters, Erin and Katie, who currently attend Michigan universities.